AF588363

Springer Theses

Recognizing Outstanding Ph.D. Research

Aims and Scope

The series "Springer Theses" brings together a selection of the very best Ph.D. theses from around the world and across the physical sciences. Nominated and endorsed by two recognized specialists, each published volume has been selected for its scientific excellence and the high impact of its contents for the pertinent field of research. For greater accessibility to non-specialists, the published versions include an extended introduction, as well as a foreword by the student's supervisor explaining the special relevance of the work for the field. As a whole, the series will provide a valuable resource both for newcomers to the research fields described, and for other scientists seeking detailed background information on special questions. Finally, it provides an accredited documentation of the valuable contributions made by today's younger generation of scientists.

Theses are accepted into the series by invited nomination only and must fulfill all of the following criteria

- They must be written in good English.
- The topic should fall within the confines of Chemistry, Physics, Earth Sciences, Engineering and related interdisciplinary fields such as Materials, Nanoscience, Chemical Engineering, Complex Systems and Biophysics.
- The work reported in the thesis must represent a significant scientific advance.
- If the thesis includes previously published material, permission to reproduce this must be gained from the respective copyright holder.
- They must have been examined and passed during the 12 months prior to nomination.
- Each thesis should include a foreword by the supervisor outlining the significance of its content.
- The theses should have a clearly defined structure including an introduction accessible to scientists not expert in that particular field.

More information about this series at http://www.springer.com/series/8790

Xiucong Bao

Study on the Cellular Regulation and Function of Lysine Malonylation, Glutarylation and Crotonylation

Doctoral Thesis accepted by
The University of Hong Kong, China

Author
Dr. Xiucong Bao
Department of Chemistry
The University of Hong Kong
Hong Kong, China

Supervisor
Prof. Xiang David Li
Department of Chemistry
The University of Hong Kong
Hong Kong, China

ISSN 2190-5053 ISSN 2190-5061 (electronic)
Springer Theses
ISBN 978-981-15-2508-7 ISBN 978-981-15-2509-4 (eBook)
https://doi.org/10.1007/978-981-15-2509-4

This Springer imprint is published by the registered company Springer Nature Singapore Pte Ltd.
The registered company address is: 152 Beach Road, #21-01/04 Gateway East, Singapore 189721, Singapore

Supervisor's Foreword

It is a great pleasure to write this foreword for the thesis of Dr. Xiucong Bao. She joined my research group in 2012 and received her Ph.D. degree in 2016. She has made outstanding achievements in her research in the field of chemical biology, as demonstrated by her excellent publications in many top chemistry and biology journals, as well as the numerous regional and international awards she has obtained.

In my laboratory, she has been trying to develop chemistry-based approaches to understand our life processes and to address issues in human health. She is particularly interested in elucidating complex regulatory mechanisms and functions of protein posttranslational modifications (PTMs, e.g., acetylation, methylation, and phosphorylation), the naturally occurring chemical modifications of proteins after their translation from DNA. PTMs regulate essentially every biological process, ranging from DNA replication and gene expression, to cell cycle control, signal transduction, and metabolism. Given these numerous links to essential cellular processes, it is not surprising that improper regulation of PTMs often leads to human diseases such as cancer. Therefore, it is of great importance to understand the roles of PTMs in normal cell biology and disease pathogenesis. However, while more than 300 different types of protein PTMs have been identified, functions of the vast majority of them remain poorly understood. This is particularly the case for those newly identified PTMs.

The thesis work of her covered important aspects of chemical epigenetics. Her first research project focused on a recently identified PTM—protein lysine malonylation. Because of its distinct chemical structure, malonylation was believed to have significant effects on its modified proteins' structure and function. However, the lack of knowledge about the cellular substrates of malonylation curbs the enthusiasm of examining the biological significance of this new PTM. To fill this knowledge gap, she decided to take chemical tools and approaches. To be honest, I expected little on the progress by this first-year Ph.D. student who just experienced a transition from an undergraduate student major in biology to a chemistry postgraduate student. However, she indeed very much surprised and impressed me. Taking only six months, she has successfully developed a chemical 'reporter' for rapid detection of malonylated proteins. The use of this chemical 'reporter' allowed us, for the first time, to 'see' malonylation dynamics (i.e., fast processes of addition and removal of the modification) in living human cells. She then took a step forward by combing her chemical 'reporter' with the state-of-the-art mass spectrometry to comprehensively profile malonylated proteins in human proteomes and finally led to the identification of 361 new protein candidates. Her outstanding research work has greatly expanded the substrate inventory of malonylation and paved the way for further study of biological significance of this newly identified PTM. This work finally resulted in an outstanding paper and cover story in *Angewandte Chemie*, a top chemistry journal, in 2013.

After completing the first project, she has focused on the study of PTMs of histone proteins, around which genomic DNA is wrapped to form chromatin. She got interested in a recently discovered histone PTM, lysine crotonylation (Kcr). Kcr was found to be enriched at active gene promoters and potential enhancers in mammalian cell genomes, suggesting that Kcr might be a 'gene-ON' mark for actively transcribed regions of chromatin. However, functional studies of this novel histone PTM have been hindered by the fact that the regulating enzymes that catalyze the addition (i.e., 'writers') and removal (i.e., 'erasers') of Kcr are unknown. The identification of such 'writers' and 'erasers' is extremely challenging because interactions between PTMs and their regulating enzymes are usually weak and transient. Therefore, it is difficult to identify these interactions using classical affinity-based biochemical approaches. To tackle this difficulty, she extended the use of a chemical proteomics approach previously developed in my laboratory to comprehensively profile 'erasers' for histone Kcr marks. Her study led to the identification of human Sirt3 as the FIRST 'eraser' enzyme for histone Kcr. She found that Sirt3 can function as a decrotonylase to regulate histone Kcr dynamics in living cells. Her work not only demonstrated the physiological significance of histone Kcr, but also helped to unravel unknown cellular mechanisms controlled by Sirt3 that has been considered solely as a deacetylase before. Her work has received considerable attention in the fields of both chemistry and chromatin biology and was published in *eLife*, a highly prestigious journal for biomedical research.

She also focused on the study of another newly identified PTM, lysine glutarylation (Kglu). She developed new chemical approaches, in combination with biochemical, biophysical, proteomics, and cell biology methods, to examine the functions and mechanisms of histone Kglu in the regulation of chromatin structure and dynamics. In brief, she combined a chemical reporter and mass spectrometry to detect and identify 27 lysine glutarylation (Kglu) marks on human core histones. The histone Kglu sites were then systematically validated. She further investigated an evolutionarily conserved Kglu mark at histone H4 lysine 91 (H4K91glu). The replacement of H4K91 by glutamate in *S. cerevisiae* to mimic glutarylation influenced chromatin structure and thereby results in defects in cell cycle progression, DNA damage repair and telomere silencing in vivo. After graduation, she continued the study using a semisynthetic H4K91glu protein, and she found that this negatively charged modification greatly destabilized nucleosome by weakening the interactions between histone H3/H4 tetramer and H2A/H2B dimers in vitro.

In human cells, she identified Sirt7 and KAT2A as a deglutarylase and glutaryl-transferase, respectively, to modulate the dynamics of H4K91glu. The Sirt7-catalyzed removal of H4K91glu was found to be tightly associated with chromatin condensation in mitosis and in response to DNA damage. Her study designates a new histone mark (Kglu) as a new regulatory mechanism for chromatin dynamics. In addition, while at least three types of negatively charged lysine acylations (i.e., malonylation, succinylation, and glutarylation) have been discovered on histones, their roles in regulation histone and chromatin biology remained largely unknown. Her work, for the first time, provides solid evidence in vitro and in vivo for whether, and how a negatively charged modification (e.g., Kglu) can affect nucleosome and chromatin structure and dynamics and thus regulate a variety of chromatin-associated processes. This work had a very high impact on the field of chromatin biology and was published on *Molecular Cell,* a top biology journal.

She is an exceptional researcher. She is brilliant, hard-working, ambitious, and productive. She has demonstrated an outstanding ability in conducting research at the interface of chemistry and biology. She was a determined and conscientious researcher who was always concerned to get the best possible data. It was her dedication that led to high-quality results and drove the project to the successful conclusions and high-impact publications. Her thesis contains exciting and topical results and concisely covers the fundamentals of chemical epigenetics. It is written in a very clear style and is accompanied by excellent figures. This will allow even a reader outside the research field to rapidly learn about the exciting topics investigated in the thesis.

Hong Kong, China
October 2019

Prof. Xiang David Li

Parts of this thesis have been published in the following journal articles:

Bao X., Wang Y., Li X., Li X. M., Liu Z., Yang T., Wong C. F., Zhang J., Hao Q. and Li X. D.*, Identification of 'erasers' for lysine crotonylated histone marks using a chemical proteomics approach. ***eLife***, 2014, 3. (Reproduced with Permissions)

Bao X., Zhao Q., Yang T., Fung Y. M. E. and Li X. D.*, A Chemical reporter for Lysine Malonylation. ***Angew***. ***Chem***. *Int. Ed.*, 2013, *52*, 4883–4886. (Reproduced with Permissions)

Bao X., Liu Z., Zhang W., Gladysz K., Fung Y. M. E., Tian G., Xiong Y., Wong J. W. H., Yuen K. W. Y. and Li X. D.* Glutarylation of Histone H4 Lysine 91 Regulates Chromatin Dynamics. ***Mol Cell***. 2019 Sep 18. p ii: S1097-2765(19) 30657-4. (Reproduced with Permissions)

Acknowledgements

My utmost gratitude and respect go to my research supervisor and life mentor, Dr. Xiang David Li. The research work described herein, and many of my life goals could not be easily achieved without his gracious support during the last four years. His constant understanding and encouragement have been a profound source of power for me to overcome difficulties. I am deeply influenced by his devotion to scientific research. Thank you for providing me a wonderful research project and guiding me to solve problems independently. The lessons I have learned from Li will carry me through my future career.

Working in Xiang David Li' group for 4 years has been a great pleasure, thanks to all wonderful group members, Tangpo Yang, Xin Li, Zheng Liu, Ying Xiong, Yizhe Wu, Xiaomeng Li, Yihang Jing, Jianwei Lin, Yiwen Cui, and Gaofei Tian. My special thanks go to Tangpo Yang and Ying Xiong for providing their synthesized invaluable probes to carry out my research study. Thank Xin Li, Xiaomeng Li, and Yiwen Cui for the proofreading of my thesis. I'm grateful to Xin Li and Zheng Liu for help in peptide synthesis and kinetics analysis. Thanks to Xiaomeng Li for sharing his experience in cellular biology and molecular biology study. I express my appreciation to Gaofei Tian and Yiwen Cui for their efforts in sample preparation. The contribution and support from my groupmates have certainly made a great difference.

Special thanks go to our collaborators Dr. Quan Hao and Dr. Karen Wing Yee Yuen for always offering helpful advice and for always being interested in what we were doing. I thank to Dr. Yi Wang for conducting crystallization analysis in the crotonylation project and Wei Zhang for sharing with me her knowledge and providing technical support on yeast genetic manipulation and analysis. I am also grateful to Dr. Yi Man Eva Fung for her expert help on mass spectrometry data analysis.

I thank the University of Hong Kong for the award of a postgraduate studentship from March 2012 to February 2016 and CRCG Conference and Travel Grants.

And last, but far from least, I express my deepest thanks to my parents, my sisters, and all my relatives, who always support me and encourage me. I would like to thank my husband. He always understands me and provides all the support I need.

Contents

Abbreviations

Ala	Alanine
AM	Acetoxymethyl
Arg	Arginine
ATP	Adenosine triphosphate
BD	Bromodomain
ChIP	Chromatin immunoprecipitation
CLASPI	Cross-linking-assisted and SILAC-based protein identification
CoA	Coenzyme A
CPS1	Carbamoyl phosphate synthetase 1
dNTP	Deoxynucleotide
FRET	Fluorescence resonance energy transfer
Gln	Glutamine
Glu	Glutamic acid
HAT	Histone acetyltransferase
HCT	Histone crotonyltransferase
HDAC	Histone deacetylase
HDCR	Histone decrotonylase
His	Histidine
HPLC	High-performance liquid chromatography
HU	Hydroxyurea
ITC	Isothermal titration calorimetry
Kac	Lysine acetylation
Kcr	Lysine crotonylation
Kglu	Lysine glutarylation
Kmal	Lysine malonylation
Ksucc	Lysine succinylation
LC-MS	Liquid chromatography–mass spectrometry
Lys	Lysine
m/z	Mass-to-charge ratio
MMS	Methyl methanesulfonate

NAD	Nicotinamide adenine dinucleotide
Pr	Propionylation
PTM	Posttranslational modifications
SAM	S-adenosylmethionine
SILAC	Stable isotope labeling by amino acids in cell culture
SPPS	Solid-phase peptide synthesis
UV	Ultraviolet
WT	Wild-type

Chapter 1
Introduction to Protein Posttranslational Modifications (PTMs)

1.1 Protein Posttranslational Modifications

In the 'central dogma' of molecular biology, genetic information stored in DNA can be transcribed into mRNAs and eventually translated into proteins to fulfill their biological functions. However, proteomes have been found to be two to three orders of magnitude more complex (>10^7 molecular species of proteins) than the encoding genomes (3×10^4 in human genomes) would predict [1]. The complexity of human proteome is dictated by posttranscriptional RNA processing and protein posttranslational modifications (PTMs). mRNA splicing takes place within the nuclei after or concurrently with transcription in order to create mature mRNA molecules that can undergo translation [2, 3]. By this way, more than 100,000 mRNA transcripts are produced. The second way to expand proteome is the covalent posttranslational modifications (PTMs) of proteins [4].

Protein PTMs are covalent chemical modifications (e.g. methylation [5], aceytylation [6] and phosphorylation [7]) occurring on amino acid residues in proteins after their biosynthesis. 15 of the 20 common amino acid side chains can undergo such diversification [1]. So far, more than 400 types of protein PTMs have been described. The reaction, to add a PTM, is an enzyme-catalyzed transfer of endogenous electrophiles onto nucleophilic side chains of amino acids in proteins. For instance, protein lysine methyltransferases catalyze methylation of lysine side chain in an *S*-adenosylmethionine (SAM)-dependent manner [8, 9] (Fig. 1.1). For methyl transfer, the targeted lysine residue must be deprotonated. To proceed, SAM and the lysine residue are firstly bound to the catalytic pocket of methyltransferases for deprotonation of the ε-group of the lysine residue. The deprotonated ε-amine then nucleophilically attacks on the electron-deficient methyl group (electrophile) of the SAM molecule. As a consequence, the methyl group is transferred onto the lysine side chain. Sharing the similar mechanisms, ATP, acetyl-CoA and NAD serve as donors for protein phosphorylation [10], acetylation [11] and ADP ribosylation [12], respectively.

X. Bao, *Study on the Cellular Regulation and Function of Lysine Malonylation, Glutarylation and Crotonylation*, Springer Theses,
https://doi.org/10.1007/978-981-15-2509-4_1

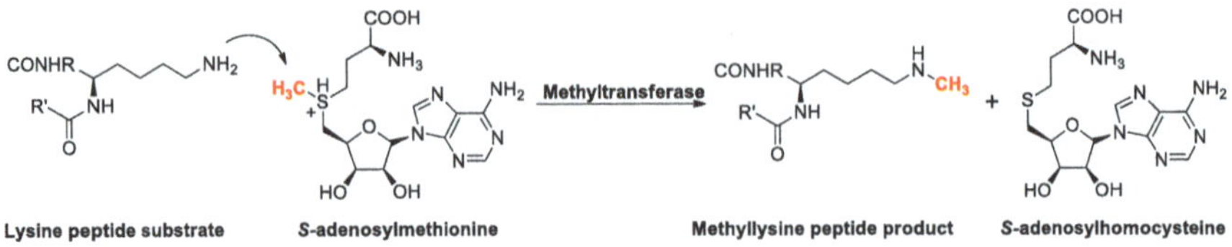

Fig. 1.1 Mechanism of lysine methylation catalyzed by histone lysine methyltransferases

PTMs play critical roles in regulating protein functions, localization and interactions in various biological processes. Protein phosphorylation is the most widespread PTM. It has been estimated that 30% of cellular proteins are targeted by phosphate group on at least one residue and that there are ~7×10^5 potential phosphorylation sites [13]. Protein phosphorylation is widely used in signal transduction. It is involved in regulation of almost all basic cellular processes, such as division, differentiation, metabolism, growth, membrane transport, motility, organelle trafficking, muscle contraction, immunity [14]. Lysine acetylation is known to regulate eukaryotic gene transcription by modifying lysine side chains of histones [15] and transcription factors [16]. In addition, over 1/3 of mitochondrial proteins are modified by acetylation [17]. The majority of acetylated proteins in mitochondria are involved in energy metabolism, linking protein acetylation to cellular metabolism. The attachment of lipid modifications, such as prenylation and palmitoylation, is involved in the change of subcellular localization of modified proteins [18]. In addition, targeted by ubiquitylation, proteins are transported to lysosome or proteasomes for proteolytic destruction [19].

1.2 Regulation of Chromatin Dynamics by Histone Modifications

The genetic information of eukaryotic cells is stored into the nucleus in the form of chromatin [20], which is made up of fundamental repeating units called nucleosomes. In nucleosome, approximately 146 bp of DNA wraps around a histone octamer consisting of two copies each of histones, H2A, H2B H3 and H4, via extensive electrostatic and hydrogen-bonding interactions [21] (Fig. 1.2). In addition to pack the large genomes to fit it into the nucleus, nucleosomes must allow appropriate DNA accessibility of proteins involved in chromatin-associated nuclear process. The structural rearrangement of nucleosomes can be achieved in several ways, such as histone replacement by the variants [22], reposition or eviction of histones by chromatin remodeling enzymes [23] and covalent histone PTMs [24]. Histone modifications have attracted attention since the discovery of association between high transcription and hyperacetylation state of histones. To this date, various histone modifications have been identified, including lysine/arginine methylation, lysine acetylation and serine/threonine phosphorylation (Fig. 1.3). Most, if not all, histone PTMs are

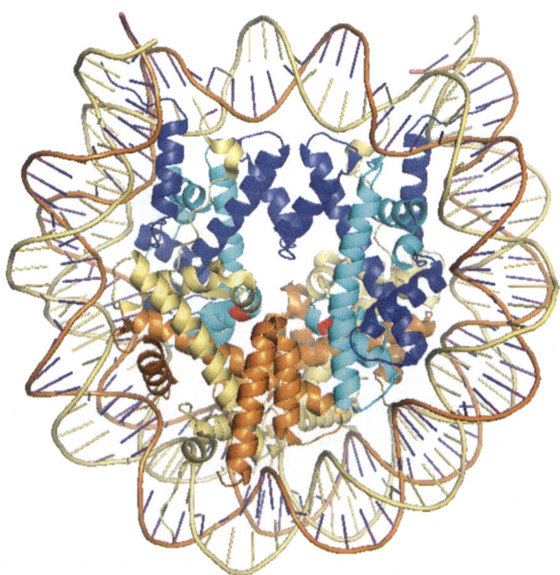

Fig. 1.2 Structure of a mononucleosome. DNA is wrapped around two copies each of H2A (yellow), H2B (orange), H3 (purple) and H4 (blue) (PDB 1KX5)

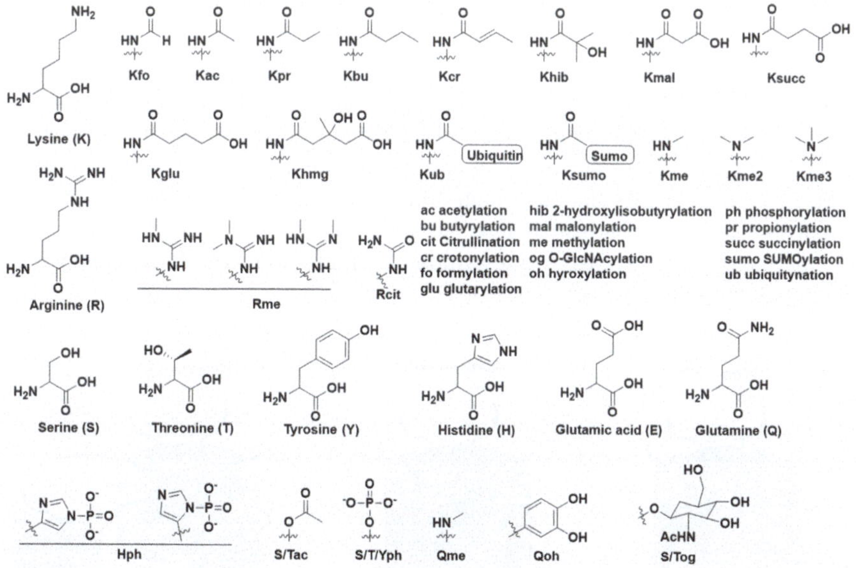

Lysine (K) Kfo Kac Kpr Kbu Kcr Khib Kmal Ksucc

Kglu Khmg Kub Ksumo Kme Kme2 Kme3

Arginine (R) Rme Rcit

ac acetylation
bu butyrylation
cit Citrullination
cr crotonylation
fo formylation
glu glutarylation
hib 2-hydroxylisobutyrylation
mal malonylation
me methylation
og O-GlcNAcylation
oh hyroxylation
ph phosphorylation
pr propionylation
succ succinylation
sumo SUMOylation
ub ubiquitynation

Serine (S) Threonine (T) Tyrosine (Y) Histidine (H) Glutamic acid (E) Glutamine (Q)

Hph S/Tac S/T/Yph Qme Qoh S/Tog

Fig. 1.3 Structures of histone post-translational modifications

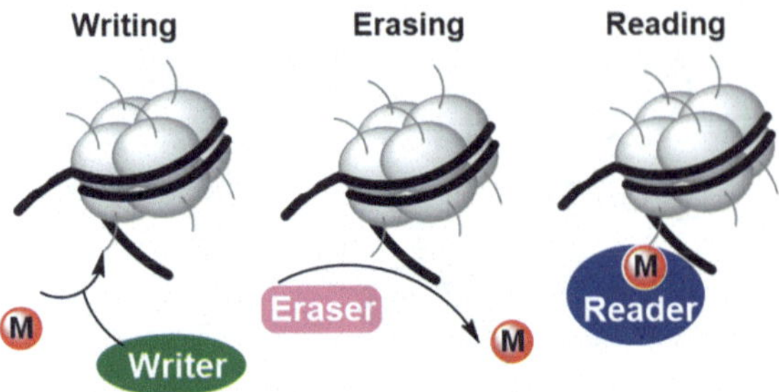

Fig. 1.4 'Writers' introduce histone marks, 'erasers' take them out and 'readers' can recognize a particular form of histone modification

reversibly regulated by a pair of opposing enzymes that catalyze the addition (so-called 'writer' enzymes) and removal (so-called 'eraser' enzymes) of the PTMs, respectively (Fig. 1.4). Taking lysine acetylation as an example, acetyltransferase [25] and deacetylase [26, 27] are responsible for 'writing' and 'erasing' acetyl group on and off lysine side chains, respectively. Histone PTMs affect chromatin structure and dynamics mainly through two ways, either directly influencing histone-DNA and histone-histone interactions or serving as docking sites to recruit effector proteins (so-called 'readers') [28] (Fig. 1.4).

There are mainly four structurally and functionally distinct regions of nucleosome, histone tails, dyad symmetry axis region, DNA entry/exist region and interfaces between histone dimers, where PTM localize can directly impact the structure of nucleosome [28] (Fig. 1.5). Histone tails, the unstructural *N*-termini of histones, extend out from the nucleosome. Removal of histone tails by trypsin has little effect on nucleosome assembly but abolishes chromatin folding, revealing the involvement of histone tails in the formation of high-order chromatin structure [29, 30]. In addition, hyperacetylation on histone tails has been found to disrupt nucleosome array folding, thereby enhance RNA transcription [31]. DNA entry/exist regions, which coordinate the outermost segments of DNA, are the first to detach from histones

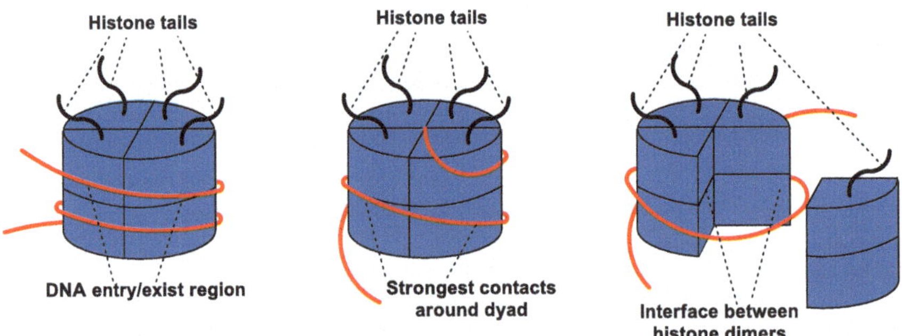

Fig. 1.5 Four structurally and functionally distinct regions of nucleosome, histone tails, dyad symmetry axis region, DNA entry/exist region and interfaces between histone dimers

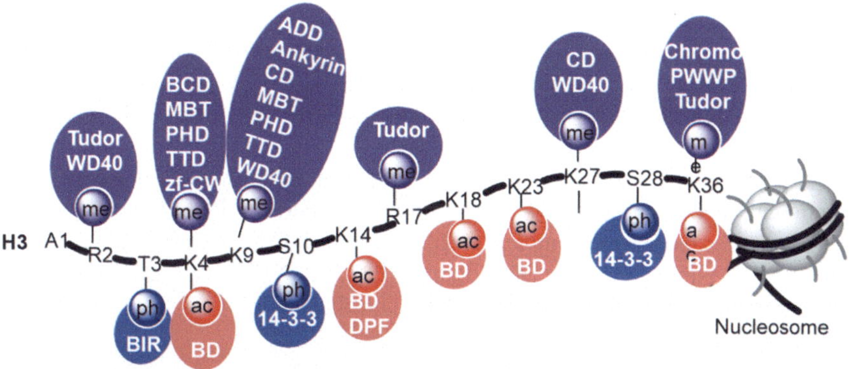

Fig. 1.6 Readers of histone PTMs

during nucleosome unwrapping. There are a number of histone PTMs described near this region involving in regulation of transcription, replication, repair and apoptosis. Specifically, acetylation at histone H3 Lys 56 (H3K56Ac) was determined, with single-molecule fluorescence resonance energy transfer (FRET), to increase the amount of DNA unwrapped that entry/exit region by 7-fold [32]. In contrast to DNA entry/exist regions, dyad symmetry axis region contains the strongest histone-DNA interactions. Acetylation on H3K122 near dyad symmetry axis was shown to facilitate DNA unwrapping and thereby cause transcription activation [33–35]. In addition to DNA-histone interactions, nucleosome stability also depends on histone-histone interactions between histone dimers. Acetylation at H4 Lys 91, which lies at the interface between H3/H4 tetramer and H2A/H2B dimers, is enriched on transcriptionally active chromatin. And mutation of Lys 91 to Ala makes nucleosomes more sensitive to micrococcal nuclease digestion [36].

In addition to direct effects on nucleosome dynamics, a set of effector proteins are recruited or repelled from chromatin, depending on the compositions of histone modifications [37] (Fig. 1.6). These proteins carry enzymatic activities to modify chromatin structure and process nuclear pathways, such as transcription, replication and repair. In 1999, the bromodomain (BD) of PCAF, which was found to recognize and bind acetylated lysine, was identified as the first reader of histone PTMs [38]. In addition, the bromodomain is also found in other factors, such as GCN5 [39], TAF1 [40] and RSC [41]. The bromodomain is the most thoroughly characterized lysine acetylation reader, which contributes to the recruitment of bromodomain-containing proteins to specific chromosomal positions wherein histones are acetylated. BD-containing proteins have also been implicated in diverse cellular processes. The bromodomain of human GCN5 is essential for transcriptional activation [39]. TAF1, which is required for the assembly of the RNA polymerase II transcription preinitiation complex, showed preference for double lysine acetylation [40]. RSC, the chromatin-remodeling complex, has 3 bromodomain containing subunits [41]. A number of 'readers' for lysine/arginine methylation have also been identified and characterized. 'Readers' of lysine methylation contain functional domains, including

ADD (ATRX-DNMT3-DNMT3L), Ankyrin, CD (chromodomain), chromo-barrel, DCD (double chromodomain), MBT (malignant brain tumor), PHD (plant home-odomain), tandem Tudor domain (TTD), Tudor, zinc-figure CW (zf-CW), WD40 and PWWP (Pro-Trp-Trp-Pro) [37]. The chromodomain of HP1 specifically recognize and bind to Lys 9 methylated histone H3 to maintain the stability of heterochromatin [42–44]. Methylation at histone H3 Lys 4 is identified as a histone mark for transcriptionally active chromatin. Recognition of this mark by PHD finger of TAF3, the subunit of transcription complex TFIID, links this PTM to transcriptional activation [45]. The binding between DCD domain of CDH1 ATPase to this PTM couples histone methylation to chromatin-remodeling mechanism [46].

1.3 Lysine Acetylation, Succinylation, Malonylation, Glutarylation and Crotonylation

1.3.1 Lysine Acetylation

1.3.1.1 Identification of Lysine Acetylation

About half a century ago, Phillips et al. and Allfrey et al. pioneered the studies on histone lysine acetylation (Kac) [6, 47] (Fig. 1.7). They discovered that the histones from cells that were incubated with radiolabeled acetate showed detectable radioactivity signals. And the incorporation of radiolabeled acetate into histones was in a translation-independent manner, demonstrating acetylation of histones as a PTM. However, due to lagging behind techniques, it had been a challenge to identify acetylation sites and acetylated proteins then. For quite a long period of time, Kac was considered as an epigenetic mark restricted to nuclei, until the identification of first non-histone acetylated protein, tubulin [48]. Subsequently, the tumor suppressor p53 [49] and HIV transcriptional activator Tat [50, 51] were identified as substrates of lysine acetylation, expanding the scope of cellular protein acetylation. Until 2006, the first reported proteomic study of protein acetylation using a specific anti-Kac antibody contributed to the identification of 338 Kac sites assigned to 195 proteins.

Fig. 1.7 The enzymatic reactions for lysine (de)acetylation

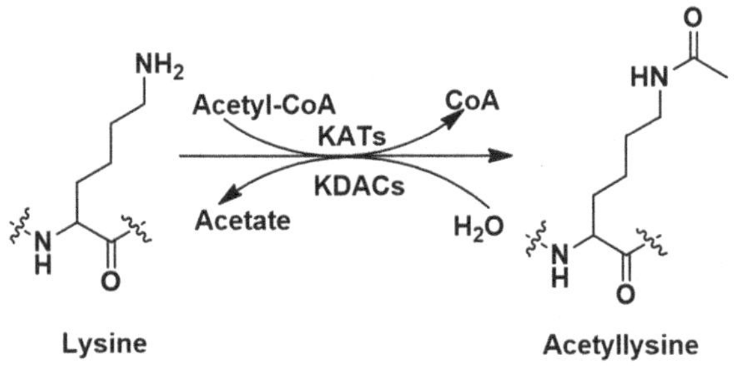

Unexpectedly, many of the acetylated proteins were reported to localize in mitochondria [52]. Three years later, taking advantage of high-resolution mass spectrometry, more acetylated substrates were uncovered, expanding the list of Kac to 3600 lysine sites on 1750 proteins with various subcellular localizations [53]. Together, protein acetylation was established as a global protein posttranslational modification.

1.3.1.2 Biological Functions of Lysine Acetylation

Bioinformatics analysis of all the identified acetylated proteins revealed that acetylated proteins are involved in various cellular pathways. Indeed, the addition of acetyl group to lysine side chain, which, on one hand, eliminates the positive charge of the ε-amino group under physiological condition, and on the other hand, introduces steric hindrance to the modified sites, has a fair chance of affecting protein structure and function. In eukaryotic cells, chromatin exist in two different states, open euchromatin and condensed heterochromatin. Histone acetylation marks are prominent at euchromatin featuring active gene transcription, suggesting the positive role of histone Kac in gene transcription (Fig. 1.8). Specifically, the absence of acetyl mark at H4 Lys 16 (H4K16Ac) is required to maintain the condensed state of heterochromatin in yeast [54] (Fig. 1.8). In addition to participation in the regulation of gene transcription, histone Kac marks have also been found to play essential roles in chromatin assembly and be associated with DNA replication and repair. In particular, to ensure its deposition onto nascent DNA sequence, histone H4 in cytoplasm is modified with multiple lysine acetylation marks at its *N*-terminal, while these marks need to be eliminated immediately after completion of nucleosome reassembly behind replication forks to maintain the stability of chromatin [55]. For another example,

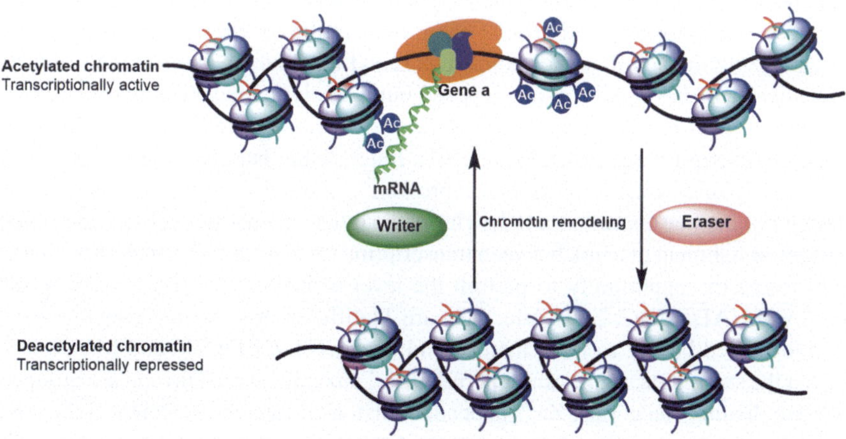

Fig. 1.8 A string of nucleosomes is shown with the tails protruding when acetylated. Histone tail acetylation results in chromatin decondensation, thereby allowing access to transcription factors and other transcription co-activators

mutation of H3K56 to Arg, which abolishes acetylation at this lysine site, results in enhanced sensitivity to genotoxic agents that interfere with DNA replication [56]. In addition, the occurrence of acetylation at histone tails destabilizes the compaction of chromatin and makes chromatin more accessible to DNA repair factors to mend the damaged DNA. Following histone lysine acetylation, several histone deacetylases (HDACs) are recruited onto chromatin near breaks to remove this mark upon completion of repair [57].

Other than the histone proteins, the non-histone acetylation has also been extensively studied. For example, the *C*-termini of tubulin undergo various modifications, which directly affect the recruitment of motor protein and microtubule-associated proteins (MAPs). By contrast, acetylation of Lys 40 on α-tubulin is an exception that its location is away from the binding sites of most motors and MAPs. The acetylation of α-tubulin is known to occur on polymerized microtubules and associated with stability of microtubules. Acetylated transcription factors, such as p53 [58] and GATA-1 [59], possess enhanced DNA-binding ability and increased transactivation functions. The interactions of HIV transcription regulator Tat with nucleic acid are also affected by acetylation. Unmodified Tat prefers to bind HIV RNA transactivating response element (TAR) and recruits pTEFb complex to promote RNA polymerase processivity, while acetylated Tat prefers to bind PCAF other than TAR to promote gene transcription elongation [60].

1.3.1.3 Dynamic Regulation of Lysine Acetylation

Protein lysine acetylation is, so far, known to play crucial roles in the regulation of a variety of cellular processes. However, it had been a long-time hypothesis that histone Kac could mediate gene transcription or degree of chromatin condensation until the discovery of Kac-modifying enzymes, histone acetyltransferses (HATs) and deacetylases (HDACs) in late 1990s. These two families of enzymes are responsible for catalyzing the addition and removal of acetyl marks to or from lysine residues, respectively. In 1996, Allis et al. screened out a catalytically active histone acetyltransferase from *Tetrahymena thermophila*, which was revealed to be orthologue of a yeast transcription regulator, GCN5 [61]. Later, Schreiber et al. identified RPD3, a previously uncovered transcription repressor, as a binding partner of trapoxin, an inhibitor of histone deacetylase [62]. Those two independent discoveries contributed to the establishment of a link between transcription regulation and histone Kac marks and offered the opportunity to perturb the level of histone Kac by regulating the activity of HAT/HDAC. Up to date, there are 22 different Kac 'writer' enzymes have been identified belong to three major families, GCN5, CBP/p300 and MYST [63, 64]. At the same time, a large amount of 'eraser' enzymes sprang up and are grouped into two distinct categories, Zn^{2+}-dependent histone deacetylases (HDAC1-11) and nicotinamide adenine dinucleotide (NAD)-dependent sirtuin deacetylases (Sirt1-7) [65, 66] (Table 1.1). Distinct from Zn^{2+}-dependent histone deacetylases, sirtuins are dependent on NAD, an important metabolic intermediate, which provides support to

Table 1.1 Subcellular localization and function of the mammalian sirtuins

Protein	Enzymatic activity	Localization	References
Sirt1	Deacetylase	Nucleus, cytoplasm	[73–76]
Sirt2	Deacetylase Demyristoylase	Nucleus, cytoplasm	[67, 77–79]
Sirt3	Deacetylase Decrotonylase	Mitochondria, nucleus	[68, 80–84]
Sirt4	ADP-ribosyltransferase	Mitochondria	[85]
Sirt5	Deacetylase Demalonylase Desuccinylase Deglutarylase	Mitochondria	[69, 70, 86–88]
Sirt6	Deacetylase ADP-ribosyltransferase Demyristoylase	Nucleus	[72, 89–93]
Sirt7	Deacetylase	Nucleolus	[94, 95]

the hypothesis that intermediary metabolites can serve as a mediator to transfer environmental change to chromatin-modifying proteins. In addition, among the seven sirtuins, Sirt3 and Sirt5 have been reported to be mostly present in mitochondria. The mitochondrial localization of HDACs inspired the identification of mitochondrial proteins as acetylation substrates. Interestingly, acetylation-independent deacylase activities have been detected in Sirt2 [67], Sirt3 [68] (see Chap. 3), Sirt5 [69–71] (see Sect 1.3 in this chapter) and Sirt6 [72]. The biological essence of dynamic regulation of lysine acetylation has been further highlighted by the discoveries of HDAC inhibitors, suberanilohydroxamic acid (SAHA), vorinostat, romidepsin and panabinostat, which have been approved for clinical treatment [27, 65, 66].

1.3.1.4 Epigenetic Readers of Lysine Acetylation

In addition to direct effect on chromatin structure, Kac can function as a docking site to recruit effector proteins, which in turn, initiate downstream signaling pathways. The speculation that bromodomain can recognize and bind to acetylated histones in chromatin was proposed as the presence of bromodomain in nuclear HATs. Indeed, Dhallui et al. demonstrated that the bromodomain of p300/CBP-associated factor (PCAF) could interact specifically with acetylated lysine, introducing the concept of 'reader' proteins to lysine acetylation field [38]. In human proteome, there are 61 bromodomains present in 46 proteins, and those proteins are involved in the regulation of various cellular pathways [96] (Table 1.2). For instance, the first identified Kac 'reader', PCAF, is a gene transcriptional coactivator [97]. Mixed-lineage leukemia (MLL) is a functional histone methyltransferase [98]. TATA-binding protein associated factors (TAFs) can serve as scaffold for assembly of transcription complex and bind to gene promoters to properly position the polymerase for gene transcription

Table 1.2 Bromodomain-containing proteins and their functions

Bromodomain-containing proteins and their functions		
Proteins	Functions	References
CREBBP, EP300, GCN5L2, PCAF	Histone acetyltransferase	[25, 38, 64]
ATAD2A/B, SMARCA2A/B, SMARCA4	ATPase	[104, 105]
BAZ1A/B, BRDT, CECR2, FALZ, WDR9	Chromatin assembly and remodeling	[106–109]
MLL, ASH1L	Methyltransferase	[98, 110, 111]
BRD1, BRD2, BRD3, BRD4, BRD7, BRD8A/B, PRKCBP1, SP100/SP110/SP140, ZMYND11, TAF1/TAF1L, TRIM24/TRIM28/RIM33/TRIM66	Transcription factor	[40, 112–114]
BRPF1A/B	MOZ complex subunit	[115]
BRWD3	JAK/STAT signaling	[116, 117]
PB1, BRD9	SWI/SNF subunit	[118, 119]
PHIP	Insulin signaling	[120, 121]
BAZ2A/B	Subunit of the nucleolar remodeling complex NoRC	[122, 123]
BRPF3A	Form a tetrameric complex with HBO1	[124]

initiation [99, 100]. The recruitment of transcription regulators to acetylated proteins may partially elucidate the functional outcomes of dynamic acetylation state. In addition, DPF3b can recognize and bind histones H3 and H4 in an acetylation-dependent manner via its PHD fingers [101, 102]. Recently, YEATS domain was also identified to specifically bind acetylation mark at histone H3 Lys 9 [103], as another alternative to bromodomain for Kac binding, suggesting there might be additional Kac 'readers' waiting to be uncovered.

1.3.2 Lysine Succinylation

1.3.2.1 Identification of Lysine Succinylation

Lysine succinylation (Ksucc) (Fig. 1.9), a covalent installation of succinyl group at the ε-amino group of lysine, was recently identified and verified as a new member of protein PTMs [125, 126]. The existence of lysine succinylation had been suggested in several pioneering works. Ron et al. initially observed a chemical group with a mass shift of 100.019 occurring on homoserin trans-succinylase (HTS). In consideration of the HTS-catalyzed reaction of transferring a succinyl group from

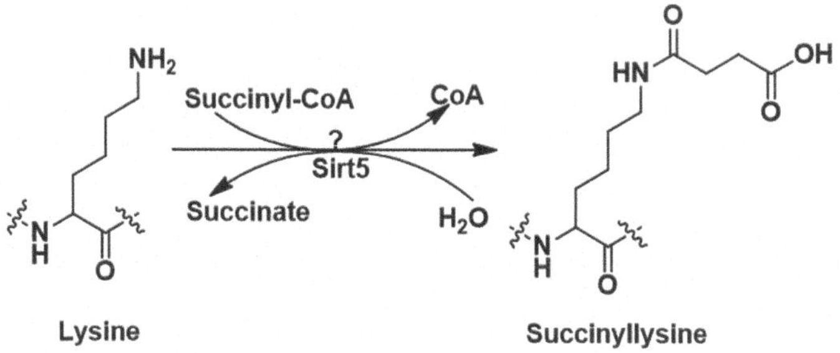

Fig. 1.9 The hypothesized enzymatic reactions for lysine (de)succinylation

succinyl-CoA to homoserine and perfect match with calculated shift of succinyl, they determined such a mass shift was caused by succinyl [127]. Osawa et al. demonstrated the form of Ksucc as a product derived from docosahexaenoic acid (DHA) in vitro and characterized the in vivo formation of Ksucc upon DHA treatment using an anti-Ksucc antibody [128]. However, both of the results suffered from insufficient evidence to distinguish Ksucc from its isomer, methylmalonyl. To conclusively identify Ksucc as a PTM, Zhao et al. focused on *E. coli* isocitrate dehydrogenase (IDH) and detected a tryptic peptide with a mass shift of 100.0186 localized at lysine residue. To ensure the observed mass shift was caused by succinyl rather than methylmalonyl, a standard lysine succinylated peptide was synthesized. Matching well with the gold standard for peptide confirmation, the synthesized peptide was revealed to have the same MS/MS spectrum and be coeluted in HPLC with the in vivo-derived peptide. The presence of succinyllysine on IDH was further confirmed by immunoblotting analysis using a specific-Ksucc antibody. In addition, isotopic succinate (2,2,3,3-D_4-succinate) labeling resulted in the detection of Ksucc-containing peptides carrying either H_4- succinyl group or D_4-succinyl group.

1.3.2.2 Succinyl-CoA as a Donor for Lysine Succinylation

Like acetyl-CoA, which serves as cofactor lysine acetylation, succinyl-CoA was proposed to be succinyl donor for lysine succinylation. Succinyl-CoA, a metabolic intermediate, is converted from α-ketoglutarate by mitochondrial α-ketoglutarate dehydrogenase complex, in which KGD1 is essential for its dehydrogenase activity [129]. Succinyl-CoA is then converted to succinate by succinyl-CoA ligase, which consists of LSC1 and LSC2. Similar to KGD1, LSC1 is also required for the activity of succinyl-CoA ligase [130]. Based on these, Choudhary et al. designed an experiment to examine the effect of varied concentration of succinyl-CoA on protein lysine succinylation via manipulation of expression of KGD1 or LSC1 [131]. Briefly, wild-type yeast cells, *lsc1Δ* and *kgd1Δ* cells were cultured in synthetic complete media supplemented with light lysine (normal), medium lysine (D_4-lysine) or heavy lysine ($^{13}C_6$,$^{15}N_2$-lysine), respectively, in the presence/absence of galactose. As expected, depletion of KGD1 led to remarkable reduction of both global and mitochondrial lysine succinylation, since loss of KGD1 blocks the conversion of α-ketoglutarate to succinyl-CoA. Meanwhile, significant increase in succinylation was observed in cells with loss of LSC1, due to the accumulation of succinyl-CoA. The presence of galactose, which is known to induce the formation of succinyl-CoA, also resulted in upregulated succinylation [130]. Consistently, the supplement of three key metabolic intermediates, glucose, pyruvate and succinate stimulated remarkable increase of lysine succinylation.

1.3.2.3 Identification of Sirt5 as a Desuccinylase

Mammals have seven sirtuin proteins (Sirt1 to Sirt7) defined as NAD-dependent deacetylase. However, only Sirt1-3 show robust deacetylation activity, whereas there is weak or even no detectable deacetylase activity of Sirt4 to Sirt7. To dig the underlying mechanism, Hao and Lin et al. carried out quantitatively enzymatic activity analysis of all seven sirtuins towards 16 different acetyl peptides and crystallization analysis of Sirt5 in complex with thioacetyl peptides [69]. The results revealed that the catalytic efficiency of Sirt5 was only ~0.2% of that of Sirt1. But sequence selectivity couldn't account for the lack of robust deacetylase activity of Sirt5, since the most hydrogen bonds adopted for interactions between thioacetyl peptide and Sirt5 were similar to a robust deacetylase, Sir2Tm. However, different from Sir2Tm complex, in which the acetyl group was surrounded by three hydrophobic residues, a larger pocket was found in Sirt5 and bound by CHES with the help of Tyr 102 and Arg 105, which showed preference for negatively charged carboxylates, such as succinyl and malonyl. Consistently, Sirt5 was demonstrated to catalyze hydrolysis of lysine succinylation on endogenous proteins, as the depletion of Sirt5 resulted in accumulation of lysine succinylation on CPS1, a known substrate of Sirt5.

1.3.2.4 Profile of Lysine Succinylated Substrates

The addition of PTM may lead to the change in physicochemical properties of crucial residues. And thus change has been revealed to affect all aspects of a protein, such as size, structure, folding state, enzymatic activity or even degradation. Lysine succinylation, which adds not only a bulky moiety with four more carbon atoms but also a terminal carboxylic acid to lysine side chain, is likely to cause more dramatic effects compared to lysine acetylation. Recently, Neill et al. detected increased succinylation level on several proteins upon lipopolysacchride (LPS) induction and identified succinate as immune signaling during inflammation, further highlighting the biological significance of Ksucc [132]. Therefore, it was urgent to profile and identify succinylated protein substrates for comprehensive characterization of the newly identified PTM. To this end, several anti-Ksucc antibodies were generated to specifically enrich succinylated peptides proteins prior to systematically mass spectrometry analysis. In combination with stable isotope labeling by amino acids in cell culture (SILAC) technology, Choudhary et al. mapped 2572, 1345, 2004 and 2140 succinylated lysine sites occurring on 990, 474, 738 and 750 proteins in *E. coli.*, *S. cereviaias*, HeLa cell and mouse liver tissue, respectively [130]. Zhao et al. identified 2580 lysine-succinylated sited on 670 proteins in *E. coli.* and 2565 succinylation sites on 779 proteins in mouse liver tissue and embryonic fibroblasts (MEFs) together [88]. Verdin et al. detected 1190 unique succinylated lysine sites across 140 proteins in mouse liver mitochondria [133]. There were 686 succinylated proteins with 1739 succinylation sites and 1545 sites on 626 proteins of *Mycobacterium tuberculosis* identified in two independent groups [134, 135]. In addition, 425 lysine succinylation sites were identified to install on 147 proteins in *Toxoplasma gondii* [136]. More interestingly, histones are also subject to Ksucc with 13, 7, 10 and 7 modified sites in HeLa, MEFs, Drosophila S2 and *S. cerevisiase* cells, respectively, indicating the potential roles of Ksucc in regulation of chromatin dynamics [137]. Taking together, those discoveries suggest Ksucc is a conserved and abundant protein covalent modification. There should be noted that even though all of those aforementioned results were obtained with the help of anti-Ksucc antibody-based affinity enrichment and even from the same samples, the profiles of succinylated proteins were different. The varied affinity and specificity of antibodies might result from the different sequence of succinylated peptides or proteins that were used as antigens.

1.3.2.5 Functional Studies of Lysine Succinylation

Among those succinylated proteins, there were two succinylated lysine residues identified in isocitrate dehydrogenase at Lys 100 and Lys 242 [125]. Even though neither of them is known to directly involve in substrate binding, the unmodified Lys 242 could form a salt bridge with near residues, Glu238 and Asp 279 and succinylated Lys 100 may also form a salt bridge with nearby Arg 119, one of the responsible residues for substrate binding. It indicated the potential role of Ksucc in regulation of enzymatic activity. In support of this argument, mutagenesis from either Lys 100 or Lys 242 to glutamate, as a mimic of lysine succinylation, decreased or completely abolished the enzymatic activity. HMGCS2 (3-hydroxy-3-methylglutaryl-CoA synthase 2), the rate-liming step of ketone body synthesis, was succinylated at 15 lysine residues [133]. Crystal structure analysis revealed that residues 83 and 310 were adjacent to the substrate-binding pocket of HMGCS2. Single or double mutations from Lys to Glu, mimicking the negative charged succinyl group, resulted in complete loss of enzymatic activity. In addition to metabolic enzymes, the essence of Ksucc on histones was highlighted since the mutation of a Ksucc site H4 Lys 31 led to negative effects on cell viability and the mutation of H3K79 or H4K77 caused gene derepression at both telomere and rDNA [137]. So far, Sirt5 is the only identified desuccinylase, catalyzing the hydrolysis of lysine succinylation, in mammalian cells. Carbamoyl phosphate synthase (CPS1) [86], a known substrate of Sirt5, was modified by succinylation at Lys 44, Lys 287 and Lys 1291 [69]. Upon Sirt5 knockout, CPS1 activity was enhanced by 15% with dramatic succinylation level increase at Lys

1291. In the same way, hypersuccinylation negatively regulated the enzymatic activities of two metabolic protein complexes, pyruvate dehydrogenase complex (PDH, catalyzing oxidation of pyruvate to acetyl-CoA) and succinate dehydrogenase (SDH, catalyzing transfer from succinate to fumarate) [88]. Furthermore, Zhao el al. found that 90% of the quantifiable Ksucc sites in MEFs showed increased abundance in Sirt5 KO cells. Out of the 1190 Ksucc sites, 64% were only detectable in cells lacking Sirt5 in Verdin's research [133]. Those discoveries together highlight the biological importance of Ksucc and essence of Sirt5. However, Sirt5 can't be the exclusive 'eraser' enzyme for Ksucc, as only part of succinylome are under the control of Sirt5. There should be other unknown regulating enzymes waiting to be unveiled.

1.3.2.6 Non-enzymatic Lysine Succinylation

Lysine acetylation is a dynamic PTM and reversibly controlled by two opposing enzymes, lysine acetyltransferases and deacetylases. Histone acetyltransferases and deacetylases-mediated changes in histone acetylation can occur over a short time, consistent with the rapid kinetics characteristic of enzymes. In contrast, mitochondria only contain one known desuccinylase, Sirt5, which globally catalyze the removal of succinyl from mitochondrial proteins, while knowledge about Ksucc 'writer' enzymes remains blank. To explain the high abundance of Ksucc in mitochondria, a nonenzymatic mechanism has been proposed [138]. Given the distinct physiological conditions of mitochondria matrix from other cellular compartments, Payne et al. demonstrated that chemical conditions in mitochondria matrix, pH value at 7.9–8.0 that may promote the extrusion of H^+ ions to generate a proton-motive force and steady-state high concentrations of metabolic intermediates, are sufficient to bring about the enzyme-independent protein acetylation and succinylation. If it was true, the lysine residues in abundant proteins would be most accessible for succinylation. As a result, more Ksucc sites would be detected on those abundant proteins. To support this hypothesis, Choudhary et al. determined the relationship between succinylation and protein abundance and confirmed this positive relevance [131]. In addition, succinylation could be detected on both overexpressed phosphoglycerate kinase (PGK) and its foreign protein tag, GFP, in similar degree. It is likely that succinylation can occur by direct chemical modification on lysine. However, it can't rule out the possibility of presence of succinyltransferase, in view of the massive existence of cytoplasmic and nuclear succinylated proteins.

1.3.3 Lysine Malonylation

1.3.3.1 Identification of Lysine Malonylation

Lysine malonylation (Kmal), a covalent modification of protein with a malonyl group incorporated at the ε-amine group of lysine (Fig. 1.10), was recently identified mainly

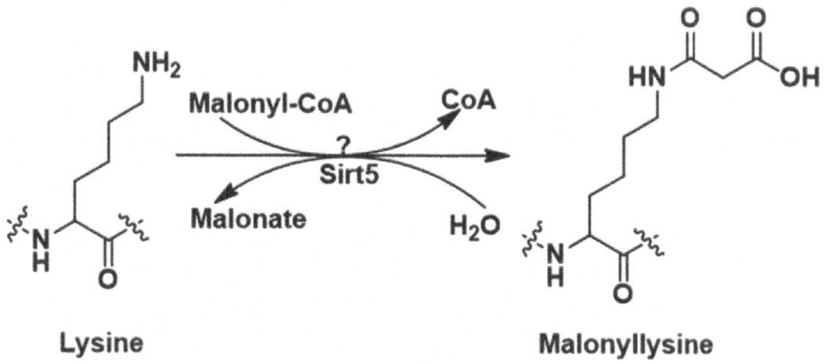

Fig. 1.10 The hypothesized enzymatic reactions for lysine (de)malonylation

through two approaches [69, 71]. One is based on systematical quantification of enzymatic activity of sirtuin family proteins and sophisticated structural analysis of Sirt5, a known weak NAD-dependent deacetylase. In another method, taking advantage of advanced mass spectrometry technology, a mass shift of 86 Da was detected and validated to be caused by malonyllysine. As mentioned above, Sirt5 prefers to bind and catalyze the removal of negatively charged carboxylates, such as malonyl group, rather than acetylated lysine. In line with this discovery, Zhao el al. performed a fluorescence-based assay to measure the demalonylation activity of all HDACs and found that only Sirt5 exhibits significant demalonylase activity in vitro. Sirt5 could also catalyze hydrolysis of lysine malonylation on endogenous proteins, as the depression and depletion of Sirt5 resulted in accumulation of lysine malonylation. Meanwhile, lysine malonylation was decreased in response to upregulated Sirt5 expression. In addition, Verdin et al. detected significant increase in malonylation level on 120 proteins in Sirt5 knockout animals [87].

To determine whether lysine malonylation exists in vivo, Hao and Lin et al. used mass spectrometry and identified two mitochondrial enzymes, glutamate dehydrogenase and malate dehydrogenase with three and two malonyllysine residues, respectively [69]. To facilitate the identification of malonylated proteins, Zhao et al. generated a specific anti-Kmal antibody to enrich Kmal peptides from a tryptic digest of both HeLa and *E. coli* whole-cell lysate and subjected them to HPLC-MS/MS analysis [71]. There were 25 peptides from 17 proteins and 3 peptides from 3 proteins from HeLa cell and *E. coli*, respectively, identified to be potential candidates of lysine malonylation. To determine whether malonyl was responsible for the observed mass shift, standard peptides bearing malonylated lysine were synthesized and analyzed by HPLC–MS/MS. If the in vivo-derived digested peptides and the synthetic peptides are indistinguishable in HPLC chromatography and share same MS/MS spectrum, they can be considered identical. By this way, the aforementioned mass shift of 86 Da was determined to result from Kmal. In addition, the stable-isotope metabolic labeling using [1,2,3-^{13}C]-malonate led to identification of peptides that were labeled with [1,2,3-^{13}C]-malonyl. It indicated that malonate could be converted into an intermediate, malonyl-CoA, which in turn was used as a donor for lysine malonylation. However, the intrinsic limitations of antibody-based approaches, potential cross-reactivity and epitope disruption or occlusion caused by presence of additional PTMs at adjacent or nearby residues, make it unideal to profile more candidate proteins. Recently, chemical reporters, as a complementary approach, have been developed to provide direct and real-time monitor of the targeting proteins, including but not limited to acetylation [139]. In this study, we developed a chemical reporter, MalAM-yne, for lysine malonylation [140]. The application of this reporter allowed the visualization and identification of malonylated protein substrates from bacteria to human cells. Furthermore, the use of chemical reporter made it possible to monitor the dynamics of lysine malonylation regulated by Sirt5 (see Chap. 2).

1.3.3.2 Malonyl-CoA an Essential Metabolic Intermediate

Malonyl-CoA, a cofactor for lysine malonylation, is an important metabolic intermediate involved in biosynthesis of fatty acid. Cellular malonyl-CoA is synthesized mainly from carboxylation of acetyl-CoA catalyzed by acetyl-CoA carboxylase (ACC) [141] and β-oxidation of old-chain-length dicarboxylic acids [142]. The consumption of malonyl-CoA is caused by malonyl-CoA decarboxylase (MCD), fatty acid elongases and synthase [143]. ACC2 deficient mice are resistant to obesity and diabetes despite consuming high calorie diets, therefore ACC2 has been proposed to be an attractive therapeutic target for obesity and diabetes [144]. Malonyl-CoA decarboxylase deficiency is an inborn metabolic disorder caused by deficient MCD activity [145]. Higher levels of malonyl-CoA are observed in type 2 diabetic patients [146]. The vital roles played by malony-CoA in metabolic homeostasis suggest the potential significance of lysine malonylation in cellular regulation in response to dynamics of cellular malonyl-CoA.

1.3.3.3 Biological Significance of Lysine Malonylation

To characterize the potential roles of lysine malonylation in human diseases, Wei et al. and Zhao et al. carried out systematic analysis of lysine malonylation patterns in different disease mouse models [147]. Given the elevated malonyl-CoA level observed in type 2 diabetic patients, Wei et al. chose a classic type 2 diabetic mouse model, *db/db* mice with point mutation in the gene for leptin receptor and investigated seven types of lysine acylations in liver tissue lysates from both wild-type and *db/db* mice. Different from lysine acetylation, monomethylation, dimethylation, succinylation, crotonylation, butyrylation and propionylation, only lysine malonylation showed marked elevation in *db/db* mices compared with wild-type ones. Consistently, enhanced lysine malonylation was also observed in another type 2 diabetic mice model, *ob/ob* mice. There were total 573 lysine sites in 268 proteins identified to be modified by malonylation via affinity enrichment using anti-Kmal antibody and subsequent mass spectrometry analysis. There were 246 lysine malonylation sites in 169 proteins overlapped between the two groups but with strengthened signal in *db/db* livers than wild-type ones. Specifically, 28 malonylated peptides were detected in

10-formyltetrahydrofolate dehydrogenase (10-FTHFDH), a one-carbon metabolic enzyme. And the malonylation at K107 on aldolase B, fructose-bisphosphate B (ALDOB) was found to modulate its enzymatic activity. In Zhao's study, two different experimental models were used: Sirt5 KO mice liver and MCD-deficient fibroblasts from malonic aciduria patients. There were 4016 malonylated lysine sites in 1395 proteins identified in livers from Sirt5 KO mice and 4943 lysine sites with malonylation in 1831 proteins in human fibroblasts. Among those identified Kmal sites and proteins in MCD human fibroblasts, malonylation level at 461 sites on 339 proteins were increased by above 2-fold and 1452 of 1762 unquantified Kmal sites were only detected in MCD deficient cells. Bioinformatics analysis of all the identified malonylated proteins reveal the involvement of lysine malonylation in regulation of various metabolic pathway, highlighting the significance of lysine malonylation in maintenance of metabolic homeostasis.

1.3.4 Lysine Glutarylation

1.3.4.1 Hypothesis of Lysine Glutarylation

Extracellular environmental changes can often lead to fluctuation of cellular energy status. To maintain the cellular metabolic homeostasis, cells have evolved various adaptive strategies, including alteration of epigenetic marks to modulate gene transcription, regulation of enzymatic activity by protein PTMs or cellular metabolites [148]. Metabolic intermediates, like acyl-CoA and SAM, have been found to function as donors for various protein PTMs, and the fluctuation of cellular NAD concentration is involved in enzymatic activity regulation of sirutins [8, 9]. In addition, lysine succinylation and malonylation have been identified as new protein PTMs [71, 126, 131], and succinyl-CoA and malonyl-CoA have been uncovered to serve as cofactors for these newly identified PTMs. Glutaryl-CoA is also an important intermediate in the metabolism of lysine and tryptophan. Furthermore, the structural similarity between glutaryl-CoA with succinyl-CoA and malonyl-CoA inspired the hypothesis that glutaryl-CoA could serve as a donor for posttranslational lysine glutarylation (Kglu) (Fig. 1.11 and 1.12).

1.3.4.2 Identification of Lysine Glutarylation

To test this hypothesis, we synthesized a chemical reporter, GluAM-yne, for the identification of lysine glutarylated proteins. As shown in Chap. 3, a wide range of proteins was labeled by this chemical reporter. At the same time, Zhao et al. generated a pan-anti-Kglu antibody [70]. The specificity of this antibody was validated via testing against peptides bearing varied modifications at lysine residue, including glutarylation, acetylation, malonylation and succinylation. Only peptide containing glutarylated lysine could be recognized by this antibody, demonstrating

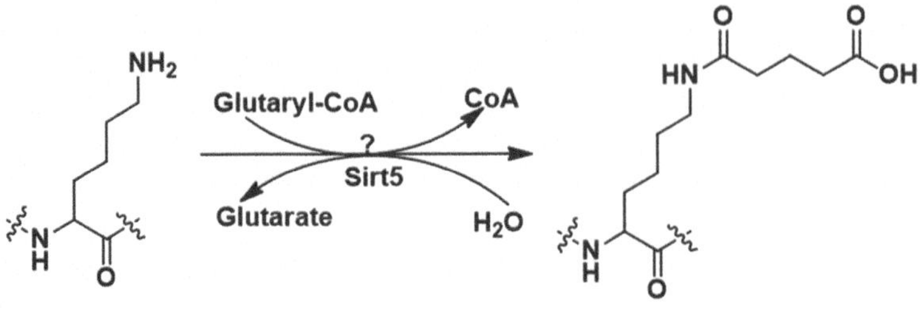

Fig. 1.11 The hypothesized enzymatic reactions for lysine (de)glutarylation

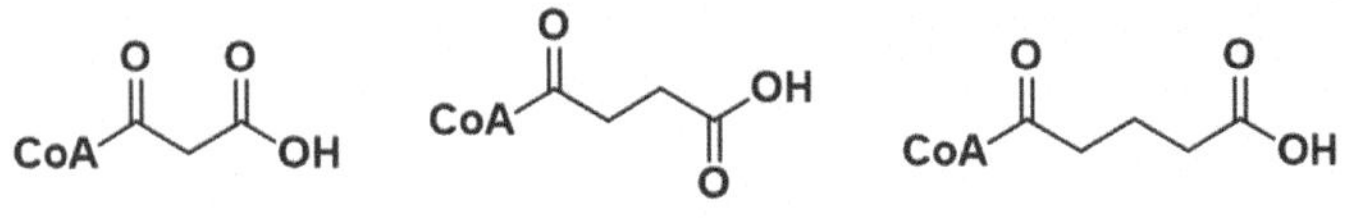

Fig. 1.12 Structures of malonyl-CoA, succinyl-CoA and glutaryl-CoA

its high selectivity. Immunoblotting analysis using this antibody against whole-cell lysate from different species—bacteria, *S. cerevisiae*, mouse and human—revealed the evolutionary conservation of Kglu. There were 23 and 10 glutarylated lysine sites identified in *E. coli* and HeLa cells, respectively, via antibody-based proteomics analysis. The confidence of Kglu as protein PTM, was enhanced by the identical MS/MS spectrum and elution time in HPLC between standard glutarylated peptides and in vivo-derived peptides with a mass shift of 114.028 at lysine site. Apart from that, the incubation with isotope D_4-glutarate resulted in detection of peptide labeled with D_4-glutaryl. More importantly, we have identified and validated lysine glutarylation as a new histone mark (see Chap. 3).

1.3.4.3 Sirt5 Identified as a Deglutarylase

Sirt5, in favor of negatively charged carboxylate, was identified an 'eraser' for lysine malonylation and succinylation. To test whether Sirt5 could also work as a deglutarylase, three different methods were carried out, fluorescence-based assay, consumption monitor of ^{32}P-radiolabeled NAD and HPLC-MS analysis against known Sirt5-targeted peptides. Among all the tested HDACs (HDAC1-11 and Sirt1-7), only Sirt5 showed NAD-dependent deglutarylation activity. The kinetic analysis revealed that the deglutarylation activity of Sirt5 was much higher than demalonylation activity, but lower than its desuccinylation activity. More importantly, Sirt5 could work as an endogenous deglutarylase, as knockout of Sirt5 (Sirt5-/-) showed significant impact on global Kglu level and proteomic analysis using liver from Sirt5-/- mice identified 911 Kglu sites on 229 proteins.

1.3.4.4 Function Studies of Lysine Glutarylation

Bioinformatics analysis of the identified glutarylated proteins revealed that Kglu was significantly enriched on proteins involving various metabolic pathways, like fatty acid and amino acid metabolism and enzymatic processes. CPS1, with 33 Kglu sites, is one of the heavily glutarylated substrates. Hyperglutarylation on CPS1 upon Sirt5 knockout led to reduced enzymatic activity. Among 44 chemically glutarylated lysine sites, 14 sites were with remarkably downregulated glutarylation level upon Sirt5 treatment. To determine whether dietary manipulation can affect protein Kglu, three different analysis were performed. Firstly, the altered Kglu level were detected in both wild-type and Sirt5-/- mice during fasting. Second, the supplementation with tryptophan, whose degradation generates glutaryl-CoA, increased the level of both glutaryl-CoA and protein glutarylation. Finally, a mouse model of human glutaric acidemia (GA), generated by mutation in gene of glutaryl-CoA decarboxylation (*GCDH*), displayed increased cellular glutaryl-CoA and protein Kglu. More importantly, glutarylation localized at H4 Lys 91 was revealed to interrupt the interaction between H3/H4 tetramer and H2A/H2B dimers, thereby result in unstable nucleosome structure (see Chap. 3).

1.3.5 Lysine Crotonylation

1.3.5.1 Identification of Lysine Crotonylation

The majority of known histone PTMs are installed within the unstructured N-terminal tails of core histones to recruit or repel chromatin binding proteins. Recently the PTMs identified in the globular domain of core histones have been found to directly modulate chromatin dynamic structure via direct interference with histone-DNA or histone-histone interactions. With the aid of unbiased mass spectrometry analysis, more novel PTMs and PTMs sites taking place outside the *N*-terminal tails continued to be discovered, typically lysine crotonylation (Kcr)[149] (Fig. 1.13). Taking advantage

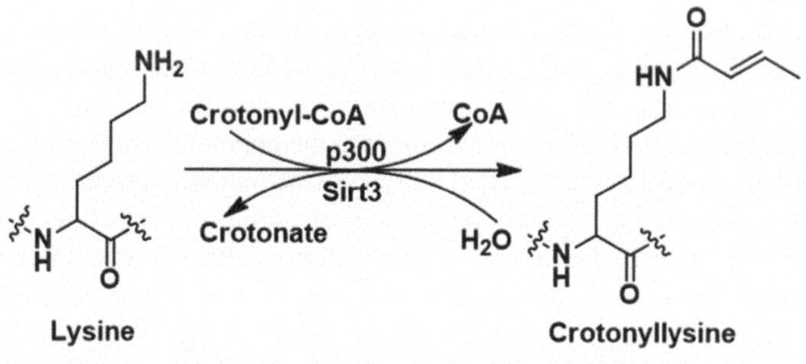

Fig. 1.13 The hypothesized enzymatic reactions for lysine (de)crotonylation

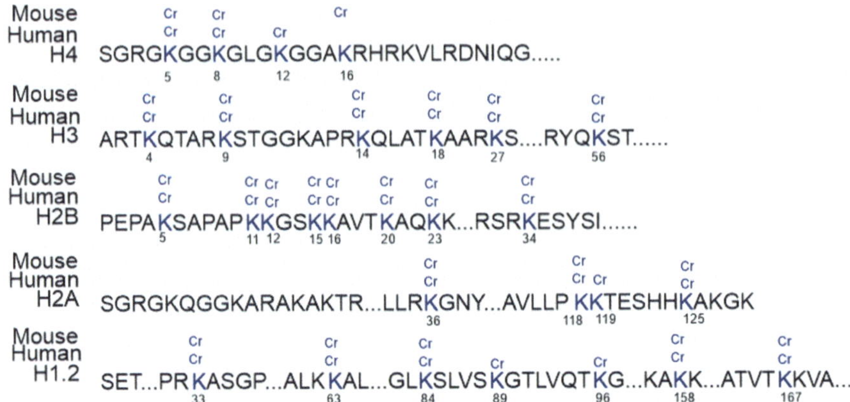

Fig. 1.14 Illustrations of histone lysine crotonylation (cr) sites in human HeLa cells and mouse MEF cells. All Kcr sites are shown in blue

of in vitro chemical derivatization, efficient peptide separation and high sensitivity of mass spectrometer, Zhao et al. successfully identified Kcr as a conserved histone PTM (Fig. 1.14). The confidence of this discovery came from several evidence, identical mass spectrum and indistinguishable HPLC chromatographic profiles with standard crotonylated peptides, immunoblotting and immunostaining analysis via specific anti-Kcr antibody and isotope labeling of D_4-crotonate.

1.3.5.2 Histone Lysine Crotonylation in Gene Transcription

Lysine crotonylation, similar to lysine acetylation, could neutralize positively charged lysine. But different form lysine acetylation, it carries a fixed structure due to C=C double bond. Therefore, it is highly possible that the presence of Kcr on histones can exhibit its functional importance, like Kac that is correlated with stimulative gene transcription, but recruit transcription factors in different way due to the structure difference. As shown in a systematic chromatin immunoprecipitation combined with DNA sequencing (ChIP-seq) analysis using pan-anti-Kcr antibody, abundance of Kcr were detected on active chromatin, like transcription starting sites (TSS), enhancers and promoters, highlighting the positive role of Kcr in gene transcription [149]. There was a significant overlap between genomic localization of Kcr and that of Kac. However, the difference emerged when investigation the genomic distribution of histone marks in spermatogenic cells [150]. The process of meiosis is known to associate with sex chromosome inactivation. After meiosis, chromosome X-linked genes are needed to be reactivated. The genomic localization analysis revealed that the post-meiotic histone Kcr marks were present on X and Y chromosome, with 31.4% and 1.2%, respectively, while histone Kac marks didn't show such preference. The dissimilarity between Kcr and Kac is further amplified by the discovery that among 49 human bromodomains, only the second bromodomain in TAF1 is capable of binding lysine crotonyl mark [151].

1.3.5.3 Regulation of Histone Lysine Crotonylation

Several questions are then raised. How a 'choice' is made between acetylation versus crotonylation? What are the regulating enzymes responsible for addition and removal of Kcr mark? Does it share the same enzymes for histone Kac dynamics? To identify histone crotonyltransferases (HCT) capable of transferring crotonyl group to targeted residues, Allis et al. carried out fractionation of nuclear extract from HeLa S3 cell to purify HCT via antibody based HCT assay [152]. By this way, several HATs were screened out as HCT candidates, they are p300, GCN5, TIP60 and MOF. Among those candidates, only p300 exhibited measurable HCT activity. Semiquantitative mass spectrometry analysis by spectral counting and immunoblotting analysis revealed and confirmed the site of both p300-catalyzed histone acetylation and crotonylation is Lys 18 on histone H3. Consistent with aforementioned preference of Kcr at active chromatin, p300-catalyzed histone crotonylation stimulated gene transcription. Here comes the question again, what is the decisive factor to help p300 make a choice between Kcr and Kac? A hypothesis was proposed that the cellular concentrations of crotonyl-CoA and acetyl-CoA make the decision. Using a well-characterized transcription program, LPS-initiated stimulation of macrophages, the upregulated gene expression and cytokine secretion have been linked to the increase of H3K18Cr mark at gene promoters. Those findings indicate that intracellular crotonyl-CoA could stimulate p300-catalyzed histone crotonylation and subsequent gene transcription. But what are the sensors of cellular acyl-CoA? How can they translate the information to p300 to make such a decision? Those questions are still unclear. And the situation has become more complicated, as we have identified Sirt3, a known robust deacetylase, as an 'eraser' for histone lysine crotonylation mark (see Chap. 3).

1.4 Approaches for Identification of Protein Substrates of Specific Modifications

The primary requirement to characterize protein PTMs is to identify their protein substrates. Recently, mass spectrometry-based proteomics have greatly contributed to the identification of hundreds of PTMs. Since protein PTMs are often substoichiometric, specific enrichment is required to detect and characterize low relative abundant components. To facilitate it, strategies based on chemical affinity, immunoaffinity and bioorthogonal chemical reporters have been developed and applied to unveil protein modifications and modifications sites.

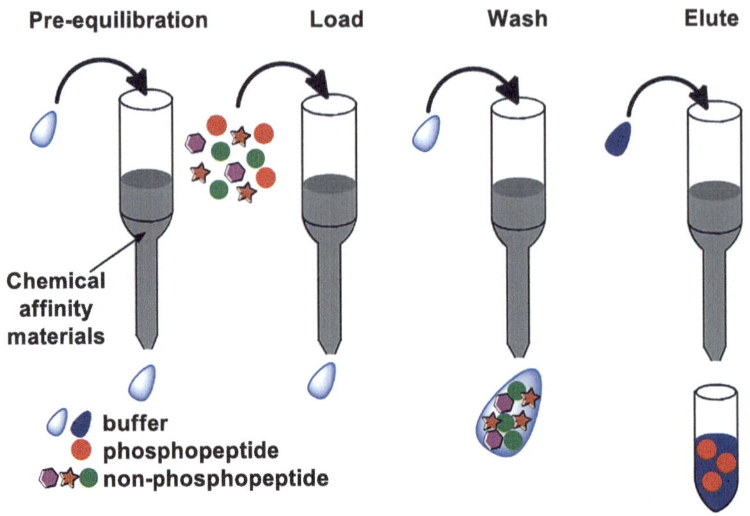

Fig. 1.15 Illustration of chemical affinity approach for enrichment of phosphorylated proteins/peptides

1.4.1 Chemical Affinity Enrichment

Protein phosphorylation are often substoichiometric and therefore the modified peptide is at such a low concentration that unmodified peptides can suppress the detection of phosphopeptides. To conquer this difficulty, the fraction of unmodified peptides relative to the PTM variants can be reduced via enrichment of modified peptides based on chemical affinity [153] (Fig. 1.15). For example, phosphopeptides can be isolated from complex mixtures of peptides utilizing a strong cation exchange resin, in which a metal chelating agent is bond to trivalent metal cations, such as Fe^{3+} or Ga^{3+}, the approach is term as IMAC [154]. There is another approach using titanium dioxide (TiO_2) as a substitute for the metal charged resin. The phosphorylated peptides can bind to TiO_2 under acidic conditions and then are eluted out by shifting the solution into basic buffer [155]. Those two methods are complementary to each other, since IMAC prefer to bind to peptides with multiple phosphorylation, while TiO_2 is characterized by higher binding affinity to singly phosphorylated peptides.

1.4.2 Antibody Based Immunoaffinity Enrichment

Immunoaffinity chromatography is another type of affinity chromatography, in which biochemical mixtures can be separated based on a specific interaction between antigen and antibody. The antibody-based immunoaffinity enrichment has been widely used for identification of proteins substrates with specific PTMs (Fig. 1.16). To do

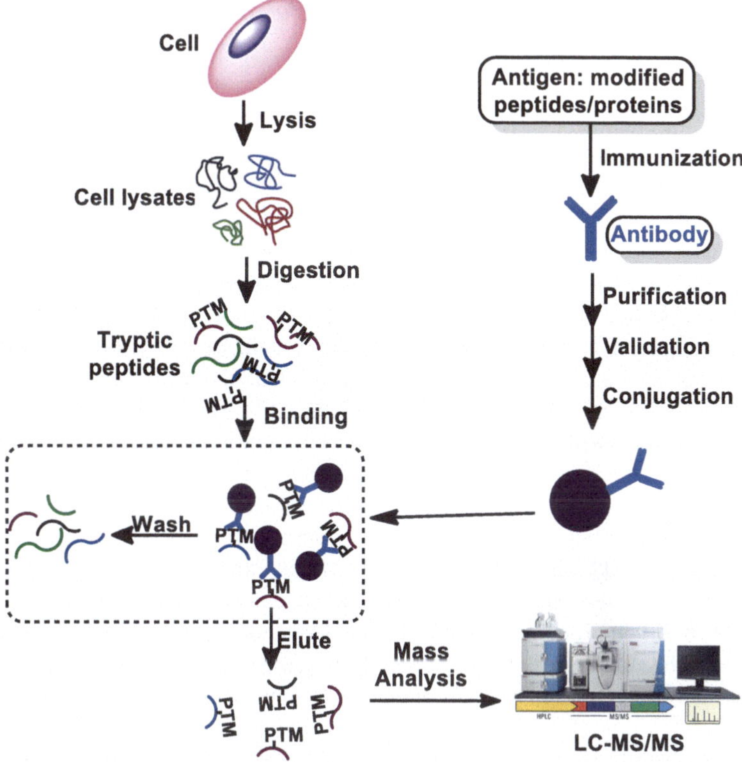

Fig. 1.16 Illustration of immunoaffinity approach for enrichment of PTM modified peptides for mass spectrometry analysis

this, peptides or proteins with chemical modifications are used as antigens to initiate the production of corresponding antibodies. And then the purified antibodies will be immobilized onto supports to isolate target proteins from complex protein mixtures for further sequence analysis. The studies utilizing PTM specific antibodies have contributed to high identification coverage and quantitative profiling of protein phosphorylation [156, 157] and methylation [158, 159]. Recently, the application of anti-malonyl/succinyl/glutaryllysine antibodies have facilitated the identification of proteins with lysine modified by malonylation [71], succinylation [125, 126] or glutarylation [70].

1.4.3 Chemical Reporters for Protein Posttranslational Modifications

Bioorthogonal chemistry refers to chemical reaction that can take place in living systems without disturbing native biochemical processes [160]. There are main two steps proceeding for the use of bioorthogonal chemistry: (i) decoration of a bioorthogonal functional group on a cellular substrate without interfering with its bioactivity (termed as chemical reporters), and the acceptance of these functionalized chemical reporters by native or engineered enzymes is the key step to this bioorthogonal labeling strategy, (ii) introduction of the chemical reporter in living system for metabolic labeling onto targeted substrates, that allows the visualization and enrichment purification of endogenous biomolecules in the aid of click chemistry reactions, such as Cu (I)-catalyzed azide-alkyne cycloaddition (Fig. 1.17). The application of chemical reporters have been widely used to and advanced the studies of various protein PTMs given their proper superiorities relative above-mentioned approaches [139]. Firstly, the demanding of affinity tags that often suffers from interference with biological activity of labeled biomolecules, has been decoupled from chemical reporters with small size and good biological compatibility of chemical tags, like azide or alkyne. Second, chemical reporters could provide direct detection of targeted substrates, like PTM-modified proteins, in a way that is independent of surrounding functional groups, which is always trouble for antibody-based on assay. Third, like radioactive tracers, chemical reporters provide another efficient method to monitor dynamics of PTMs, but it can also enable more sensitive detection and direct identification of labeled substrates. Finally, bioorthogonal chemical has facilitated the analysis of interactions between small molecules and proteins, beneficial to biochemical target identification.

The presence of lipid modifications of proteins has been revealed to affect the cellular localization, membrane affinity and message courier roles in cell signaling of the targeted proteins [161]. The detection of lipidated proteins is, therefore, required to decipher biological role of cellular lipidation. However, even though some forms of protein lipidation can be predicted based on the conserved amino acid motifs, there is still lacking direct biochemical way. Unlike soluble proteins, the hydrophobic nature of lipidated proteins, especially membrane proteins, makes traditional methods difficult to implement. The deficiency of lipidation-specific antibodies further exacerbates the situation. The development of bioorthogonal chemical

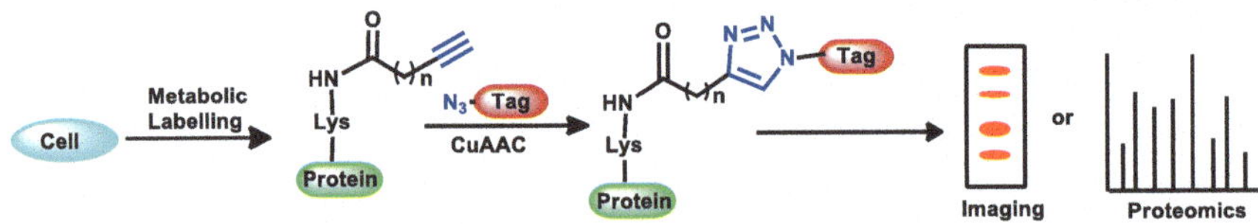

Fig. 1.17 Illustration of chemical reporter-based approach for the detection and identification of PTM modified proteins

reporters bursts the study of lipidation via providing a valuable way for discovery and characterization of lipidated proteins [162]. Protein *S*-prenylation is a prominent form of protein lipidation, for which, the use of azide [163] or alkyne [164, 165] isoprenoid reporters enabled the large-scale profiling of known and unexpected modified proteins. For myristoylation of *N*-terminal glycine residues (*N*-myristoylation), the application of myristoylation chemical reporters contributed to the identification of IPAJ, a bacterial secreted proteolytic enzyme working on *N*-myristoylated host proteins to benefit the virulence of bacterial pathogen [166]. The combination between metabolic labeling of alkyne fatty acid reporters and bioorthogonal reaction also enabled the visualization of detection of *S*-palmitoylated proteins [162]. Recently, the bifunctional fatty acid chemical reporter, x-alk-16, was demonstrated to not only incorporate into known modified proteins, but also capture *S*-palmiotyl mediated interaction proteins via covalent cross-linking [167].

In addition, chemical reporters have been widely used for investigation of protein PTMs, including glycosylation [168], which plays a crucial role in inside protein translocation, signal transduction and extracellular protein interaction, ADP-riosylation [169] that represents one of mechanisms for bacterial virulence, AMPylation [170], used as an infection strategy by bacteria, and various acylations, such as acetylation [171], malonylation [140].

1.5 Identification of Protein-Protein Interactions

As aforementioned one essential mechanism adopted by histone lysine acetylation to mediate chromatin structure as well as associated nuclear processes is to serve as a signal that initiates the recruitment of regulatory proteins onto chromatin. Therefore, it is essential to probe PTM-mediated protein-protein interactions for comprehensive exploration of PTM-involved biological functions, especially for those newly identified PTMs. In this section, the typical approaches used for identification of protein-protein interactions will be discussed.

1.5.1 Yeast 2-Hybrid System

The yeast two-hybrid system is a genetic system to detect the interaction between two proteins of interest via the reconstitution of a transcription factor, upon which the subsequent reporter genes can be activated [172] (Fig. 1.18). Briefly, transcription factor is split into two independent functional domains, DNA binding domain (DBD) and activation domain (AD), and there are two proteins of interest fused to each of the functional domain (DBD-X as bait and AD-Y as prey). DBD can bind to promoter element upstream of a reporter gene without activation, while AD is capable of activing the gene transcription but has no affinity to the operator sequences. Only the occurrence of interaction between proteins X and Y, can DBD and AD fuse to

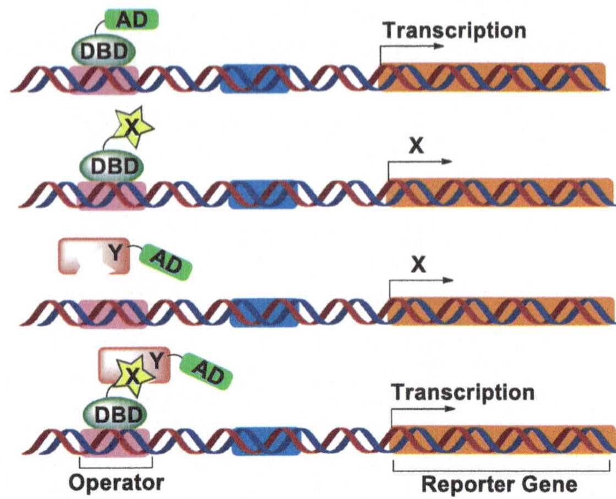

Fig. 1.18 Illustration of yeast 2-hybrid system for identification of protein-protein interactions

reconstitute a functional transcription factor to activate reporter gene. For large-scale screenings, there are two approaches: the library screening approach and the matrix approach. In library screening approach, a particular bait is transformed into type-a yeast strain and a batch of preys is expressed in type-α yeast. After mating between bait-carrying strain and mixture of library strains, only clones expressing interaction pair of proteins can survive on selective media. And then the corresponding plasmids will be isolated and sequenced to determine the identity of the interacting prey. While in matrix approach, a series of defined interactions will be assayed instead of screening of a collection of unknown preys.

1.5.2 Protein Affinity Chromatography

Protein affinity chromatography is taking advantage of the resistance of highly bonded protein mixture to high salt treatment (Fig. 1.19). To do this, a target protein can be covalently coupled onto resin or agarose and then the mixture of extracted proteins will pass over a column containing immobilized protein. Most proteins with little-to-no affinity to the target protein will pass through the columns or are easily washed off by low-salt buffer, whereas high affinity binding proteins will retain on column and then be eluted by high-salt solutions or even protein detergent, such as sodium dodecyl sulfate (SDS). Those strongly retained fractions will be subjected to mass spectrometry to identify the binding partners. As a technique for detection of protein-protein interaction, there is a distinct advantage of protein affinity chromatography, that it is easy to determine functional domains and critical residues within a protein for the specific interaction. However, this system can lead to false-negative results in the case that the interacting protein may not be able to interact with the target protein once it is fastened on resin or false-positive results due to the indirect bonding.

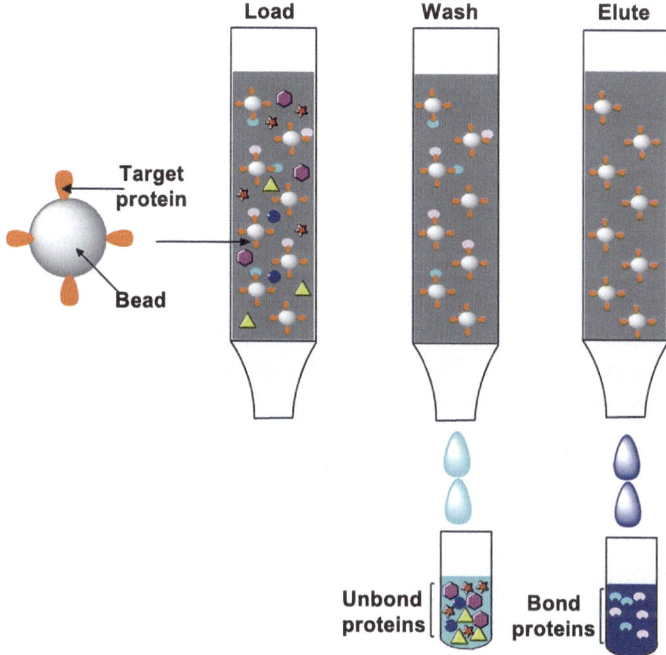

Fig. 1.19 Illustration of protein affinity chromatography for identification of protein-protein interactions

1.5.3 Co-immunoprecipitation (Co-IP)

Co-immunoprecipitation, a classical method making use of the specific interaction between antibody and antigen, has been widely used for the investigation of protein-protein interactions [173] (Fig. 1.20). The fundamental principle of the assay is

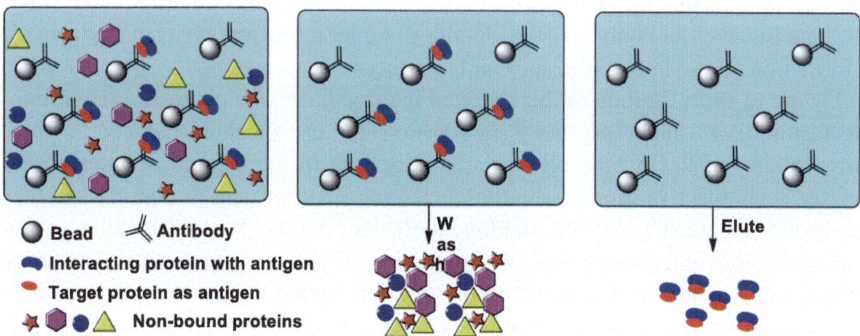

Fig. 1.20 Illustration of coimmunoprecipitation method for identification of protein-protein interactions

quite simple, that is, the prey proteins forming mixture with bait protein will be specific extracted by anti-bait antibody. Simply, anti-bait antibody is fastened on magnetic/agarose beads before or after addition into whole-cell lysate to precipitate antigen-containing complex. Unbound proteins will be washed away and isolated proteins will be analyzed via immunoblotting or mass spectrometry to confirm the identity of prey proteins. Immunoprecipitation is a simple and powerful method to explore protein-protein interaction. However, it will not be an option once the specific antibody is unavailable. Alternatively, the fusion of an epitope tag with commercially available antibodies onto overexpressed bait protein can be used to solve the problem. Like protein affinity chromatography, immunoprecipitation can detect the interactions in the presence of all the competing proteins in a crude lysate. Different from protein affinity chromatography, immunoprecipitation is able to detect in *in vivo* protein-protein interactions. However, the low concentration of antigen can lead to less sensitivity of the approach. To deal with it, excessive antigen can be added into the crude lysates to drive complex formation.

1.5.4 Photo-Cross-Linking

The traditional approaches mentioned above for protein-protein identification rely on either indirect signal detection (i.e. reporter gene transcription) or immunoprecipitation of protein complex of interest. The adoption of those methods can highly possible result in false-positives. In the case of yeast 2-hybrid system, protein fragments that interact nonspecifically with baits or proteins that can directly bind to operator sequence upstream of the reporter gene while having no affinity toward bait protein, can also activate reporter gene transcription. For immunoaffinity-based systems, the indirect interaction can't be excluded thoroughly, and the weak interaction proteins can be washed away in harsh condition. To overcome those difficulties, photo-induced cross linking-based proteomics approach has been developed for identification of protein–protein interactions. The method, which can capture interactors by converting the weak and transient interactions into much more stable covalent linkages, not only give the identities of interaction proteins, but also provide information on the interaction sites and interfaces.

Owing to short lifetime of their excited intermediates, the photo-reactive groups are featured with high specificity. In addition, the use of photo-induced covalent cross-linking makes it possible to temporally control the capture of interactions. Briefly, a photoactive group is firstly labeled onto bait protein or peptide and then any proteins that can recognize and bind to the bait will be fixed by forming a covalent bond upon light (usually by UV) stimulation. To do this, photoactivable function groups can be site-specifically incorporated into peptide using solid-phase peptide synthesis (SPPS), and the installation of photo-cross-linking groups into proteins can be achieved via semi-synthesis (i.e. native chemical ligation) or biochemical unnatural amino acids (UAAs) incorporation using suppressor tRNAs for translation in vitro and in vivo. To facilitate the isolation of captured proteins, Li et al.

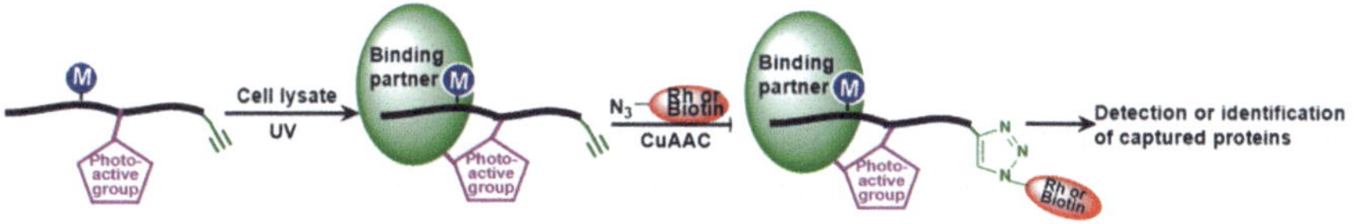

Fig. 1.21 Photo-cross-liking strategy to capture proteins recognizing histone PTMs

developed a chemical approach to profile PTM-mediated protein interactions [174] (Fig. 1.21). In this method, in addition to incorporation of photoactive group capable of converting non-covalent binding into covalent bond to catch transient interaction, a functional group (i.e. alkyne) has been appended onto probe to facilitate the conjugation of reporter tags for visualization or affinity enrichment of interactors. In his design, a peptide probe was synthesized based on *N*-terminal sequence of histone H3 with Lys 4 trimethylated. A photo-cross-linker (benzophenone) was appended to Ala 7, mostly close to the modified site but without interfering the protein-protein interaction. In addition, an alkyne-containing amino acid was added at the *C*-terminus. He firstly tested the ability of the newly designed chemical reporters to covalently label PHD finger of ING2, a known 'reader' for trimethylated lysine. Upon UV stimulation, the recombinant protein was fixed to peptide probe. The labeled protein was then conjugated to rhodamine-azide via Cu(I)-catalyzed azide-alkyne cycloaddition. The robust and specific labelling of ING2 by H3K4me3 probes was visualized via in-gel fluorescence imaging. The capability of benzophenone-based chemical reporters in identifying histone PMTs specific binding proteins was further confirmed by successful capture of known recombinant (ING2, BPTF and JMJD2A) or endogenous (ING2) H3K4me3 binding proteins in complex proteomes. To profile 'readers' for H3K4me3, Li et al. combined the photo-cross-linking strategy with SILAC (Stable Isotope Labeling by Amino acids in Cell culture)-based quantitative mass spectrometry, termed CLASPI (Cross-Linking-Assisted and SILAC-based Protein Identification) [175]. Using this method, MORC3, in addition to aforementioned H3K4me3 'readers' and SPIN1, was for the first time revealed to specific recognize histone mark, H3K4me3. The CLASPI strategy is also applicable to reveal histone H3T3phos readers and also contributes to unveil Sirt3 as an 'eraser' for histone lysine crotonylation mark [68] (see Chap. 3).

1.6 The Objectives and Organization of This Thesis

The initial goal of this research was to decipher protein posttranslational modifications via multiple approaches with respect to profile of protein substrates of lysine malonylation by alkyne-functionalized chemical reporters (Chap. 2), identification of lysine glutarylation as new histone mark (Chap. 3), investigation the role of H4K91 glutarylation in modulation of nucleosome assembly (Chap. 4) and characterization of Sirt3 as an 'eraser' for histone lysine crotonylation mark (Chap. 5).

References

1. Walsh CT, Garneau-Tsodikova S, Gatto GJ, Jr. (2005) Protein posttranslational modifications: the chemistry of proteome diversifications. Angewandte Chem 44:7342–7372. https://doi.org/10.1002/anie.200501023
2. Black DL (2003) Mechanisms of alternative pre-messenger RNA splicing. Annu Rev Biochem 72:291–336. https://doi.org/10.1146/annurev.biochem.72.121801.161720
3. Maniatis T, Tasic B (2002) Alternative pre-mRNA splicing and proteome expansion in metazoans. Nature 418:236–243. https://doi.org/10.1038/418236a
4. Walsh C (2006) Posttranslational modification of proteins: expanding nature's inventory. Roberts and Company Publishers, Englewood, CO
5. Greer EL, Shi Y (2012) Histone methylation: a dynamic mark in health, disease and inheritance Nature reviews. Genetics 13:343–357 https://doi.org/10.1038/nrg3173
6. Allfrey VG, Faulkner R, Mirsky AE (1964) Acetylation and methylation of histones and their possible role in the regulation of rna synthesis. Proc Natl Acad Sci USA 51:786–794. https://doi.org/10.1073/pnas.51.5.786
7. Mann M, Ong SE, Gronborg M, Steen H, Jensen ON, Pandey A (2002) Analysis of protein phosphorylation using mass spectrometry: deciphering the phosphoproteome. Trends Biotechnol 20:261–268
8. Paik WK, Paik DC, Kim S (2007) Historical review: the field of protein methylation. Trends Biochem Sci 32:146–152. https://doi.org/10.1016/j.tibs.2007.01.006
9. Smith BC, Denu JM (2009) Chemical mechanisms of histone lysine and arginine modifications. Biochem Biophys Acta 1789:45–57. https://doi.org/10.1016/j.bbagrm.2008.06.005
10. Adams JA (2001) Kinetic and catalytic mechanisms of protein kinases. Chem Rev 101:2271–2290
11. Jiang J et al (2012) Investigation of the acetylation mechanism by GCN5 histone acetyltransferase PloS One 7:e36660. https://doi.org/10.1371/journal.pone.0036660
12. Okazaki IJ, Moss J (1996) Mono-ADP-ribosylation: a reversible posttranslational modification of proteins. Adv Pharmacol 35:247–280
13. Cohen P (2000) The regulation of protein function by multisite phosphorylation–a 25 year update. Trends Biochem Sci 25:596–601
14. Ubersax JA, Ferrell JE Jr (2007) Mechanisms of specificity in protein phosphorylation. Nat Rev Mol Cell Biol 8:530–541. https://doi.org/10.1038/nrm2203
15. Eberharter A, Becker PB (2002) Histone acetylation: a switch between repressive and permissive chromatin. Second Rev Ser Chromatin Dyn EMBO Rep 3:224–229. https://doi.org/10.1093/embo-reports/kvf053
16. Soutoglou E, Katrakili N, Talianidis I (2000) Acetylation regulates transcription factor activity at multiple levels. Mol Cell 5:745–751
17. Anderson KA, Hirschey MD (2012) Mitochondrial protein acetylation regulates metabolism. Essays Biochem 52:23–35. https://doi.org/10.1042/bse0520023
18. Resh MD (2013) Covalent lipid modifications of proteins. Curr Biol CB 23:R431–R435. https://doi.org/10.1016/j.cub.2013.04.024
19. Hochstrasser M (1996) Ubiquitin-dependent protein degradation. Ann Rev Genet 30:405–439. https://doi.org/10.1146/annurev.genet.30.1.405
20. Kouzarides T (2007) Chromatin modifications and their function. Cell 128:693–705. https://doi.org/10.1016/j.cell.2007.02.005
21. Tan S, Davey CA (2011) Nucleosome structural studies. Curr Opin Struct Biol 21:128–136. https://doi.org/10.1016/j.sbi.2010.11.006
22. Sarma K, Reinberg D (2005) Histone variants meet their match. Nat Rev Mol Cell Biol 6:139–149. https://doi.org/10.1038/nrm1567
23. Gutierrez JL, Chandy M, Carrozza MJ, Workman JL (2007) Activation domains drive nucleosome eviction by SWI/SNF. EMBO J 26:730–740. https://doi.org/10.1038/sj.emboj.7601524

24. Huang H, Lin S, Garcia BA, Zhao Y (2015) Quantitative proteomic analysis of histone modifications. Chem Rev 115:2376–2418. https://doi.org/10.1021/cr500491u
25. Roth SY, Denu JM, Allis CD (2001) Histone acetyltransferases. Ann Rev Biochem 70:81–120. https://doi.org/10.1146/annurev.biochem.70.1.81
26. Sauve AA, Wolberger C, Schramm VL, Boeke JD (2006) The biochemistry of sirtuins. Annu Rev Biochem 75:435–465. https://doi.org/10.1146/annurev.biochem.74.082803.133500
27. Falkenberg KJ, Johnstone RW (2014) Histone deacetylases and their inhibitors in cancer, neurological diseases and immune disorders. Nat Rev Drug Disc 13:673–691. https://doi.org/10.1038/nrd4360
28. Bowman GD, Poirier MG (2015) Post-translational modifications of histones that influence nucleosome dynamics. Chem Rev 115:2274–2295. https://doi.org/10.1021/cr500350x
29. Allan J, Harborne N, Rau DC, Gould H (1982) Participation of core histone "tails" in the stabilization of the chromatin solenoid. J Cell Biol 93:285–297
30. Fletcher TM, Hansen JC (1995) Core histone tail domains mediate oligonucleosome folding and nucleosomal DNA organization through distinct molecular mechanisms. J Biol Chem 270:25359–25362
31. Simpson RT (1978) Structure of chromatin containing extensively acetylated H3 and H4. Cell 13:691–699
32. Neumann H et al (2009) A method for genetically installing site-specific acetylation in recombinant histones defines the effects of H3 K56 acetylation. Mol Cell 36:153–163. https://doi.org/10.1016/j.molcel.2009.07.027
33. Tropberger P et al (2013) Regulation of transcription through acetylation of H3K122 on the lateral surface of the histone octamer. Cell 152:859–872 https://doi.org/10.1016/j.cell.2013.01.032
34. Manohar M et al (2009) Acetylation of histone H3 at the nucleosome dyad alters DNA-histone binding. J Biol Chem 284:23312–23321. https://doi.org/10.1074/jbc.m109.003202
35. Chatterjee N et al (2015) Histone acetylation near the nucleosome dyad axis enhances nucleosome disassembly by RSC and SWI/SNF. Mol Cell Biol 35:4083–4092. https://doi.org/10.1128/mcb.00441-15
36. Ye J et al (2005) Histone H4 lysine 91 acetylation a core domain modification associated with chromatin assembly. Mol Cell 18:123–130. https://doi.org/10.1016/j.molcel.2005.02.031
37. Musselman CA, Lalonde ME, Cote J, Kutateladze TG (2012) Perceiving the epigenetic landscape through histone readers. Nat Struct Mol Biol 19:1218–1227. https://doi.org/10.1038/nsmb.2436
38. Dhalluin C, Carlson JE, Zeng L, He C, Aggarwal AK, Zhou MM (1999) Structure and ligand of a histone acetyltransferase bromodomain. Nature 399:491–496. https://doi.org/10.1038/20974
39. Syntichaki P, Topalidou I, Thireos G (2000) The Gcn5 bromodomain co-ordinates nucleosome remodelling. Nature 404:414–417. https://doi.org/10.1038/35006136
40. Jacobson RH, Ladurner AG, King DS, Tjian R (2000) Structure and function of a human TAFII250 double bromodomain module. Science 288:1422–1425
41. Kasten M, Szerlong H, Erdjument-Bromage H, Tempst P, Werner M, Cairns BR (2004) Tandem bromodomains in the chromatin remodeler RSC recognize acetylated histone H3 Lys14. EMBO J 23:1348–1359. https://doi.org/10.1038/sj.emboj.7600143
42. Nielsen PR et al (2002) Structure of the HP1 chromodomain bound to histone H3 methylated at lysine 9. Nature 416:103–107. https://doi.org/10.1038/nature722
43. Lachner M, O'Carroll D, Rea S, Mechtler K, Jenuwein T (2001) Methylation of histone H3 lysine 9 creates a binding site for HP1 proteins. Nature 410:116–120. https://doi.org/10.1038/35065132
44. Bannister AJ, Zegerman P, Partridge JF, Miska EA, Thomas JO, Allshire RC, Kouzarides T (2001) Selective recognition of methylated lysine 9 on histone H3 by the HP1 chromo domain. Nature 410:120–124. https://doi.org/10.1038/35065138
45. Vermeulen M et al (2007) Selective anchoring of TFIID to nucleosomes by trimethylation of histone H3 lysine 4. Cell 131:58–69. https://doi.org/10.1016/j.cell.2007.08.016

46. Flanagan JF et al (2005) Double chromodomains cooperate to recognize the methylated histone H3 tail. Nature 438:1181–1185. https://doi.org/10.1038/nature04290
47. Phillips DM (1963) The presence of acetyl groups of histones. Biochem J 87:258–263
48. L'Hernault SW, Rosenbaum JL (1985) Chlamydomonas alpha-tubulin is posttranslationally modified by acetylation on the epsilon-amino group of a lysine. Biochemistry 24:473–478
49. Gu W, Roeder RG (1997) Activation of p 53 sequence-specific DNA binding by acetylation of the p53 C-terminal domain Cell 90:595–606
50. Kiernan RE et al (1999) HIV-1 tat transcriptional activity is regulated by acetylation. EMBO J 18:6106–6118. https://doi.org/10.1093/emboj/18.21.6106
51. Ott M, Schnolzer M, Garnica J, Fischle W, Emiliani S, Rackwitz HR, Verdin E (1999) Acetylation of the HIV-1 Tat protein by p 300 is important for its transcriptional activity. Curr Biol CB 9:1489–1492
52. Kim SC et al (2006) Substrate and functional diversity of lysine acetylation revealed by a proteomics survey. Mol Cell 23:607–618. https://doi.org/10.1016/j.molcel.2006.06.026
53. Choudhary C et al (2009) Lysine acetylation targets protein complexes and co-regulates major cellular functions. Science 325:834–840. https://doi.org/10.1126/science.1175371
54. Millar CB, Kurdistani SK, Grunstein M (2004) Acetylation of yeast histone H4 lysine 16: a switch for protein interactions in heterochromatin and euchromatin. Cold Spring Harb Symp Quant Biol 69:193–200. https://doi.org/10.1101/sqb.2004.69.193
55. Ge Z, Nair D, Guan X, Rastogi N, Freitas MA, Parthun MR (2013) Sites of acetylation on newly synthesized histone H4 are required for chromatin assembly and DNA damage response signaling. Mol Cell Biol 33:3286–3298. https://doi.org/10.1128/mcb.00460-13
56. Yuan J, Pu MT, Zhang ZG, Lou ZK (2009) Histone H3-K56 acetylation is important for genomic stability in mammals. Cell Cycle 8:1747–1753. https://doi.org/10.4161/cc.8.11.8620
57. Tamburini BA, Tyler JK (2005) Localized histone acetylation and deacetylation triggered by the homologous recombination pathway of double-strand DNA repair. Mol Cell Biol 25:4903–4913. https://doi.org/10.1128/MCB.25.12.4903-4913.2005
58. Luo J, Li M, Tang Y, Laszkowska M, Roeder RG, Gu W (2004) Acetylation of p 53 augments its site-specific DNA binding both in vitro and in vivo. Proc Natl Acad Sci U S A 101:2259–2264
59. Lamonica JM, Vakoc CR, Blobel GA (2006) Acetylation of GATA-1 is required for chromatin occupancy. Blood 108:3736–3738. https://doi.org/10.1182/blood-2006-07-032847
60. Ott M et al. (2004) Tat acetylation: a regulatory switch between early and late phases in HIV transcription elongation. Novartis Found Symp 259:182–193; discussion 193–186, 223–185
61. Brownell JE, Zhou J, Ranalli T, Kobayashi R, Edmondson DG, Roth SY, Allis CD (1996) Tetrahymena histone acetyltransferase A: a homolog to yeast Gcn5p linking histone acetylation to gene activation. Cell 84:843–851
62. Massuda ES et al (1997) Regulated expression of the diphtheria toxin A chain by a tumor-specific chimeric transcription factor results in selective toxicity for alveolar rhabdomyosarcoma cells. Proc Natl Acad Sci USA 94:14701–14706
63. Marmorstein R (2001) Structure and function of histone acetyltransferases. Cell Mol Life Sci CMLS 58:693–703
64. Marmorstein R, Roth SY (2001) Histone acetyltransferases: function, structure, and catalysis. Curr Opin Genet Dev 11:155–161
65. Dokmanovic M, Clarke C, Marks PA (2007) Histone deacetylase inhibitors: overview and perspectives. Mol Cancer Res MCR 5:981–989. https://doi.org/10.1158/1541-7786.MCR-07-0324
66. Marks PA, Xu WS (2009) Histone deacetylase inhibitors: potential in cancer therapy. J Cell Biochem 107:600–608. https://doi.org/10.1002/jcb.22185
67. Liu Z, Yang T, Li X, Peng T, Hang HC, Li XD (2015) Integrative chemical biology approaches for identification and characterization of "erasers" for fatty-acid-acylated lysine residues within proteins. Angew Chem 54:1149–1152. https://doi.org/10.1002/anie.201408763
68. Bao X et al (2014) Identification of 'erasers' for lysine crotonylated histone marks using a chemical proteomics approach eLife 3. https://doi.org/10.7554/elife.02999

69. Du J et al (2011) Sirt5 is a NAD-dependent protein lysine demalonylase and desuccinylase Science 334:806–809. https://doi.org/10.1126/science.1207861
70. Tan M et al (2014) Lysine glutarylation is a protein posttranslational modification regulated by SIRT5. Cell Metab 19:605–617. https://doi.org/10.1016/j.cmet.2014.03.014
71. Peng C et al (2011) The first identification of lysine malonylation substrates and its regulatory enzyme. Mol Cell Proteomics MCP 10(M111):012658. https://doi.org/10.1074/mcp.M111.012658
72. Jiang H et al (2013) SIRT6 regulates TNF-alpha secretion through hydrolysis of long-chain fatty acyl lysine. Nature 496:110–113. https://doi.org/10.1038/nature12038
73. Ryall JG et al (2015) The NAD(+)-dependent SIRT1 deacetylase translates a metabolic switch into regulatory epigenetics in skeletal muscle stem cells. Cell Stem Cell 16:171–183. https://doi.org/10.1016/j.stem.2014.12.004
74. Nakahata Y et al (2008) The NAD +-dependent deacetylase SIRT1 modulates CLOCK-mediated chromatin remodeling and circadian control. Cell 134:329–340. https://doi.org/10.1016/j.cell.2008.07.002
75. Vaziri H et al (2001) hSIR2(SIRT1) functions as an NAD-dependent p 53 deacetylase. Cell 107:149–159
76. Tanno M, Sakamoto J, Miura T, Shimamoto K, Horio Y (2007) Nucleocytoplasmic shuttling of the NAD+-dependent histone deacetylase SIRT1. J Biol Chem 282:6823–6832. https://doi.org/10.1074/jbc.m609554200
77. North BJ, Marshall BL, Borra MT, Denu JM, Verdin E (2003) The human Sir2 ortholog, SIRT2, is an NAD+-dependent tubulin deacetylase. Mol Cell 11:437–444
78. North BJ, Verdin E (2007) Interphase nucleo-cytoplasmic shuttling and localization of SIRT2 during mitosis. PloS one 2:e784 https://doi.org/10.1371/journal.pone.0000784
79. Teng YB et al (2015) Efficient demyristoylase activity of SIRT2 revealed by kinetic and structural studies. Sci Rep 5:8529. https://doi.org/10.1038/srep08529
80. Hirschey MD, Shimazu T, Huang JY, Schwer B, Verdin E (2011) SIRT3 regulates mitochondrial protein acetylation and intermediary metabolism. Cold Spring Harb Symp Quant Biol 76:267–277. https://doi.org/10.1101/sqb.2011.76.010850
81. Hirschey MD et al (2010) SIRT3 regulates mitochondrial fatty-acid oxidation by reversible enzyme deacetylation. Nature 464:121–125. https://doi.org/10.1038/nature08778
82. Mahlknecht U, Voelter-Mahlknecht S (2011) Genomic organization and localization of the NAD-dependent histone deacetylase gene sirtuin 3 (Sirt3) in the mouse. Int J Oncol 38:813–822. https://doi.org/10.3892/ijo.2010.872
83. Scher MB, Vaquero A, Reinberg D (2007) SirT3 is a nuclear NAD+-dependent histone deacetylase that translocates to the mitochondria upon cellular stress. Genes Dev 21:920–928. https://doi.org/10.1101/gad.1527307
84. Iwahara T, Bonasio R, Narendra V, Reinberg D (2012) SIRT3 functions in the nucleus in the control of stress-related gene expression. Mol Cell Biol 32:5022–5034. https://doi.org/10.1128/mcb.00822-12
85. Ahuja N et al (2007) Regulation of insulin secretion by SIRT4, a mitochondrial ADP-ribosyltransferase. J Biol Chem 282:33583–33592. https://doi.org/10.1074/jbc.m705488200
86. Nakagawa T, Lomb DJ, Haigis MC, Guarente L (2009) SIRT5 Deacetylates carbamoyl phosphate synthetase 1 and regulates the urea cycle. Cell 137:560–570. https://doi.org/10.1016/j.cell.2009.02.026
87. Nishida Y et al (2015) SIRT5 regulates both cytosolic and mitochondrial protein malonylation with glycolysis as a major target. Mol Cell 59:321–332. https://doi.org/10.1016/j.molcel.2015.05.022
88. Park J et al (2013) SIRT5-mediated lysine desuccinylation impacts diverse metabolic pathways. Mol Cell 50:919–930. https://doi.org/10.1016/j.molcel.2013.06.001
89. Van Meter M, Kashyap M, Rezazadeh S, Geneva AJ, Morello TD, Seluanov A, Gorbunova V (2014) SIRT6 represses LINE1 retrotransposons by ribosylating KAP1 but this repression fails with stress and age. Nat Commun 5:5011. https://doi.org/10.1038/ncomms6011

90. Mao Z et al (2011) SIRT6 promotes DNA repair under stress by activating PARP1. Science 332:1443–1446. https://doi.org/10.1126/science.1202723
91. Liszt G, Ford E, Kurtev M, Guarente L (2005) Mouse Sir2 homolog SIRT6 is a nuclear ADP-ribosyltransferase. J Biol Chem 280:21313–21320. https://doi.org/10.1074/jbc.m413296200
92. Kawahara TL et al (2009) SIRT6 links histone H3 lysine 9 deacetylation to NF-kappaB-dependent gene expression and organismal life span. Cell 136:62–74. https://doi.org/10.1016/j.cell.2008.10.052
93. Michishita E et al (2008) SIRT6 is a histone H3 lysine 9 deacetylase that modulates telomeric chromatin. Nat 452:492–496. https://doi.org/10.1038/nature06736
94. Barber MF et al (2012) SIRT7 links H3K18 deacetylation to maintenance of oncogenic transformation. Nat 487:114–118. https://doi.org/10.1038/nature11043
95. Kiran S, Chatterjee N, Singh S, Kaul SC, Wadhwa R, Ramakrishna G (2013) Intracellular distribution of human SIRT7 and mapping of the nuclear/nucleolar localization signal. FEBS J 280:3451–3466. https://doi.org/10.1111/febs.12346
96. Filippakopoulos P, Knapp S (2014) Targeting bromodomains: epigenetic readers of lysine acetylation. Nat Rev Drug Disc 13:337–356. https://doi.org/10.1038/nrd4286
97. Cherasse Y et al (2007) The p 300/CBP-associated factor (PCAF) is a cofactor of ATF4 for amino acid-regulated transcription of CHOP. Nucleic Acids Res 35:5954–5965. https://doi.org/10.1093/nar/gkm642
98. Cao F et al (2014) Targeting MLL1 H3K4 methyltransferase activity in mixed-lineage leukemia. Mol Cell 53:247–261. https://doi.org/10.1016/j.molcel.2013.12.001
99. Meyers RE, Sharp PA (1993) TATA-binding protein and associated factors in polymerase II and polymerase III transcription. Mol Cell Biol 13:7953–7960
100. Wu SY, Chiang CM (2001) TATA-binding protein-associated factors enhance the recruitment of RNA polymerase II by transcriptional activators. J Biol Chem 276:34235–34243. https://doi.org/10.1074/jbc.M102463200
101. Lange M et al (2008) Regulation of muscle development by DPF3, a novel histone acetylation and methylation reader of the BAF chromatin remodeling complex. Genes Dev 22:2370–2384. https://doi.org/10.1101/gad.471408
102. Zeng L, Zhang Q, Li S, Plotnikov AN, Walsh MJ, Zhou MM (2010) Mechanism and regulation of acetylated histone binding by the tandem PHD finger of DPF3b. Nature 466:258–262. https://doi.org/10.1038/nature09139
103. Li Y et al (2014) AF9 YEATS domain links histone acetylation to DOT1L-mediated H3K79 methylation. Cell 159:558–571. https://doi.org/10.1016/j.cell.2014.09.049
104. Boussouar F, Jamshidikia M, Morozumi Y, Rousseaux S, Khochbin S (2013) Malignant genome reprogramming by ATAD2. Biochim Biophys Acta 1829:1010–1014. https://doi.org/10.1016/j.bbagrm.2013.06.003
105. Vangamudi B et al (2015) The SMARCA2/4 ATPase domain surpasses the bromodomain as a drug target in SWI/SNF-mutant cancers: insights from cDNA rescue and PFI-3 inhibitor studies. Cancer Res 75:3865–3878. https://doi.org/10.1158/0008-5472.can-14-3798
106. Moriniere J et al (2009) Cooperative binding of two acetylation marks on a histone tail by a single bromodomain. Nature 461:664–668. https://doi.org/10.1038/nature08397
107. Cavellan E, Asp P, Percipalle P, Farrants AK (2006) The WSTF-SNF2h chromatin remodeling complex interacts with several nuclear proteins in transcription. J Biol Chem 281:16264–16271. https://doi.org/10.1074/jbc.m600233200
108. Fairbridge NA, Dawe CE, Niri FH, Kooistra MK, King-Jones K, McDermid HE (2010) Cecr2 mutations causing exencephaly trigger misregulation of mesenchymal/ectodermal transcription factors Birth defects research Part A. Clin Mol Teratol 88:619–625. https://doi.org/10.1002/bdra.20695
109. Huang H, Rambaldi I, Daniels E, Featherstone M (2003) Expression of the Wdr9 gene and protein products during mouse development developmental. Dyn Official Publ Am Assoc Anatomists 227:608–614. https://doi.org/10.1002/dvdy.10344
110. An S, Yeo KJ, Jeon YH, Song JJ (2011) Crystal structure of the human histone methyltransferase ASH1L catalytic domain and its implications for the regulatory mechanism. J Biol Chem 286:8369–8374. https://doi.org/10.1074/jbc.m110.203380

111. Gregory GD et al (2007) Mammalian ASH1L is a histone methyltransferase that occupies the transcribed region of active genes. Mol Cell Biol 27:8466–8479. https://doi.org/10.1128/mcb.00993-07
112. LeRoy G, Rickards B, Flint SJ (2008) The double bromodomain proteins Brd2 and Brd3 couple histone acetylation to transcription. Mol Cell 30:51–60. https://doi.org/10.1016/j.molcel.2008.01.018
113. Venturini L et al (1999) TIF1gamma, a novel member of the transcriptional intermediary factor 1 family. Oncogene 18:1209–1217. https://doi.org/10.1038/sj.onc.1202655
114. Sanchez R, Zhou MM (2009) The role of human bromodomains in chromatin biology and gene transcription. Curr Opin Drug Disc Dev 12:659–665
115. Hibiya K, Katsumoto T, Kondo T, Kitabayashi I, Kudo A (2009) Brpf1, a subunit of the MOZ histone acetyl transferase complex, maintains expression of anterior and posterior Hox genes for proper patterning of craniofacial and caudal skeletons. Dev Biol 329:176–190. https://doi.org/10.1016/j.ydbio.2009.02.021
116. Muller P, Kuttenkeuler D, Gesellchen V, Zeidler MP, Boutros M (2005) Identification of JAK/STAT signalling components by genome-wide RNA interference. Nature 436:871–875. https://doi.org/10.1038/nature03869
117. Field M et al (2007) Mutations in the BRWD3 gene cause X-linked mental retardation associated with macrocephaly. Am J Hum Genet 81:367–374. https://doi.org/10.1086/520677
118. Kadoch C, Hargreaves DC, Hodges C, Elias L, Ho L, Ranish J, Crabtree GR (2013) Proteomic and bioinformatic analysis of mammalian SWI/SNF complexes identifies extensive roles in human malignancy. Nat Genet 45:592–601. https://doi.org/10.1038/ng.2628
119. Reisman D, Glaros S, Thompson EA (2009) The SWI/SNF complex and cancer. Oncogene 28:1653–1668. https://doi.org/10.1038/onc.2009.4
120. Kaburagi Y et al (2007) Role of IRS and PHIP on insulin-induced tyrosine phosphorylation and distribution of IRS proteins. Cell Struct Function 32:69–78
121. Farhang-Fallah J, Yin X, Trentin G, Cheng AM, Rozakis-Adcock M (2000) Cloning and characterization of PHIP, a novel insulin receptor substrate-1 pleckstrin homology domain interacting protein. J Biol Chem 275:40492–40497. https://doi.org/10.1074/jbc.c000611200
122. Zhou Y, Santoro R, Grummt I (2002) The chromatin remodeling complex NoRC targets HDAC1 to the ribosomal gene promoter and represses RNA polymerase I transcription. EMBO J 21:4632–4640
123. Santoro R, Li J, Grummt I (2002) The nucleolar remodeling complex NoRC mediates heterochromatin formation and silencing of ribosomal gene transcription. Nat Genet 32:393–396. https://doi.org/10.1038/ng1010
124. Yan K et al (2016) The chromatin regulator BRPF3 preferentially activates the HBO1 acetyltransferase but is dispensable for mouse development and survival. J Biol Chem 291:2647–2663. https://doi.org/10.1074/jbc.m115.703041
125. Zhang Z, Tan M, Xie Z, Dai L, Chen Y, Zhao Y (2011) Identification of lysine succinylation as a new post-translational modification. Nat Chem Biol 7:58–63. https://doi.org/10.1038/nchembio.495
126. Colak G et al (2013) Identification of lysine succinylation substrates and the succinylation regulatory enzyme CobB in *Escherichia coli*. Mol Cell Proteomics MCP 12:3509–3520. https://doi.org/10.1074/mcp.M113.031567
127. Rosen R, Becher D, Buttner K, Biran D, Hecker M, Ron EZ (2004) Probing the active site of homoserine trans-succinylase. FEBS Lett 577:386–392. https://doi.org/10.1016/j.febslet.2004.10.037
128. Kawai Y, Fujii H, Okada M, Tsuchie Y, Uchida K, Osawa T (2006) Formation of Nepsilon-(succinyl)lysine in vivo: a novel marker for docosahexaenoic acid-derived protein modification. J Lipid Res 47:1386–1398. https://doi.org/10.1194/jlr.m600091-jlr200
129. Repetto B, Tzagoloff A (1989) Structure and regulation of KGD1, the structural gene for yeast alpha-ketoglutarate dehydrogenase. Mol Cell Biol 9:2695–2705
130. Przybyla-Zawislak B, Dennis RA, Zakharkin SO, McCammon MT (1998) Genes of succinyl-CoA ligase from Saccharomyces cerevisiae. Eur J Biochem/FEBS 258:736–743

131. Weinert BT, Scholz C, Wagner SA, Iesmantavicius V, Su D, Daniel JA, Choudhary C (2013) Lysine succinylation is a frequently occurring modification in prokaryotes and eukaryotes and extensively overlaps with acetylation. Cell Rep 4:842–851. https://doi.org/10.1016/j.celrep.2013.07.024
132. Tannahill GM et al (2013) Succinate is an inflammatory signal that induces IL-1beta through HIF-1alpha. Nature 496:238–242. https://doi.org/10.1038/nature11986
133. Rardin MJ et al (2013) SIRT5 regulates the mitochondrial lysine succinylome and metabolic networks. Cell Metab 18:920–933. https://doi.org/10.1016/j.cmet.2013.11.013
134. Xie L et al (2015) First succinyl-proteome profiling of extensively drug-resistant Mycobacterium tuberculosis revealed involvement of succinylation in cellular physiology. J Proteome Res 14:107–119. https://doi.org/10.1021/pr500859a
135. Yang M et al (2015) Succinylome analysis reveals the involvement of lysine succinylation in metabolism in pathogenic Mycobacterium tuberculosis. Mol Cell Proteomics MCP 14:796–811. https://doi.org/10.1074/mcp.M114.045922
136. Li X et al (2014) Systematic identification of the lysine succinylation in the protozoan parasite *Toxoplasma gondii*. J Proteome Res 13:6087–6095. https://doi.org/10.1021/pr500992r
137. Xie Z et al (2012) Lysine succinylation and lysine malonylation in histones. Mol Cell Proteomics MCP 11:100–107. https://doi.org/10.1074/mcp.M111.015875
138. Wagner GR, Payne RM (2013) Widespread and enzyme-independent Nepsilon-acetylation and Nepsilon-succinylation of proteins in the chemical conditions of the mitochondrial matrix. J Biol Chem 288:29036–29045. https://doi.org/10.1074/jbc.M113.486753
139. Grammel M, Hang HC (2013) Chemical reporters for biological discovery. Nat Chem Biol 9:475–484. https://doi.org/10.1038/nchembio.1296
140. Bao X, Zhao Q, Yang T, Fung YM, Li XD (2013) A chemical probe for lysine malonylation. Angew Chem 52:4883–4886. https://doi.org/10.1002/anie.201300252
141. Berg JM, Tymoczko JL, Stryer L (2002) Acetyl coenzyme A carboxylase plays a key role in controlling fatty acid metabolism, 5th edn. Biochemistry, New York
142. Gaertig J, Cruz MA, Bowen J, Gu L, Pennock DG, Gorovsky MA (1995) Acetylation of lysine 40 in alpha-tubulin is not essential in Tetrahymena thermophila. J Cell Biol 129:1301–1310
143. Saggerson D (2008) Malonyl-CoA, a key signaling molecule in mammalian cells. Ann Rev Nutr 28:253–272. https://doi.org/10.1146/annurev.nutr.28.061807.155434
144. Abu-Elheiga L, Oh W, Kordari P, Wakil SJ (2003) Acetyl-CoA carboxylase 2 mutant mice are protected against obesity and diabetes induced by high-fat/high-carbohydrate diets. Proc Natl Acad Sci U S A 100:10207–10212. https://doi.org/10.1073/pnas.1733877100
145. de Wit MC et al (2006) Brain abnormalities in a case of malonyl-CoA decarboxylase deficiency. Mol Genet Metab 87:102–106. https://doi.org/10.1016/j.ymgme.2005.09.009
146. Du Y et al (2015) Lysine malonylation is elevated in type 2 diabetic mouse models and enriched in metabolic associated proteins. Mol Cell Proteomics MCP 14:227–236. https://doi.org/10.1074/mcp.m114.041947
147. Colak G et al (2015) Proteomic and biochemical studies of lysine malonylation suggest Its malonic aciduria-associated regulatory role in mitochondrial function and fatty Acid oxidation. Mol Cell Proteomics MCP 14:3056–3071. https://doi.org/10.1074/mcp.M115.048850
148. Lu C, Thompson CB (2012) Metabolic regulation of epigenetics. Cell metabolism 16:9–17. https://doi.org/10.1016/j.cmet.2012.06.001
149. Tan M et al (2011) Identification of 67 histone marks and histone lysine crotonylation as a new type of histone modification. Cell 146:1016–1028. https://doi.org/10.1016/j.cell.2011.08.008
150. Montellier E, Rousseaux S, Zhao Y, Khochbin S (2012) Histone crotonylation specifically marks the haploid male germ cell gene expression program: post-meiotic male-specific gene expression. BioEssays News Rev Mol Cell Dev Biol 34:187–193. https://doi.org/10.1002/bies.201100141
151. Flynn EM, Huang OW, Poy F, Oppikofer M, Bellon SF, Tang Y, Cochran AG (2015) A subset of human bromodomains recognizes butyryllysine and crotonyllysine histone peptide modifications. Structure 23:1801–1814. https://doi.org/10.1016/j.str.2015.08.004

152. Sabari BR et al (2015) Intracellular crotonyl-CoA stimulates transcription through p 300-catalyzed histone crotonylation. Mol Cell 58:203–215. https://doi.org/10.1016/j.molcel.2015.02.029
153. Wang ZG, Lv N, Bi WZ, Zhang JL, Ni JZ (2015) Development of the affinity materials for phosphorylated proteins/peptides enrichment in phosphoproteomics analysis. ACS applied materials & interfaces 7:8377–8392. https://doi.org/10.1021/acsami.5b01254
154. Andersson L, Porath J (1986) Isolation of phosphoproteins by immobilized metal (Fe^{3+}) affinity chromatography. Anal Biochem 154:250–254
155. Schmidt A, Csaszar E, Ammerer G, Mechtler K (2008) Enhanced detection and identification of multiply phosphorylated peptides using TiO_2 enrichment in combination with MALDI TOF/TOF MS. Proteomics 8:4577–4592. https://doi.org/10.1002/pmic.200800279
156. Rush J et al (2005) Immunoaffinity profiling of tyrosine phosphorylation in cancer cells. Nat Biotechnol 23:94–101. https://doi.org/10.1038/nbt1046
157. Bergstrom Lind S et al (2008) Immunoaffinity enrichments followed by mass spectrometric detection for studying global protein tyrosine phosphorylation. J Proteome Res 7:2897–2910. https://doi.org/10.1021/pr8000546
158. Carlson SM, Gozani O (2014) Emerging technologies to map the protein methylome. J Mol Biol 426:3350–3362. https://doi.org/10.1016/j.jmb.2014.04.024
159. Geoghegan V, Guo A, Trudgian D, Thomas B, Acuto O (2015) Comprehensive identification of arginine methylation in primary T cells reveals regulatory roles in cell signalling. Nat Commun 6:6758. https://doi.org/10.1038/ncomms7758
160. Sletten EM, Bertozzi CR (2009) Bioorthogonal chemistry: fishing for selectivity in a sea of functionality. Angew Chem 48:6974–6998. https://doi.org/10.1002/anie.200900942
161. Resh MD (2006) Trafficking and signaling by fatty-acylated and prenylated proteins. Nat Chem Biol 2:584–590. https://doi.org/10.1038/nchembio834
162. Hang HC, Wilson JP, Charron G (2011) Bioorthogonal chemical reporters for analyzing protein lipidation and lipid trafficking. Acc Chem Res 44:699–708. https://doi.org/10.1021/ar200063v
163. Kho Y et al (2004) A tagging-via-substrate technology for detection and proteomics of farnesylated proteins. Proc Natl Acad Sci USA 101:12479–12484. https://doi.org/10.1073/pnas.0403413101
164. Charron G, Tsou LK, Maguire W, Yount JS, Hang HC (2011) Alkynyl-farnesol reporters for detection of protein S-prenylation in cells. Mol BioSyst 7:67–73. https://doi.org/10.1039/c0mb00183j
165. DeGraw AJ, Palsuledesai C, Ochocki JD, Dozier JK, Lenevich S, Rashidian M, Distefano MD (2010) Evaluation of alkyne-modified isoprenoids as chemical reporters of protein prenylation. Chem Biol Drug Des 76:460–471. https://doi.org/10.1111/j.1747-0285.2010.01037.x
166. Burnaevskiy N et al (2013) Proteolytic elimination of N-myristoyl modifications by the Shigella virulence factor IpaJ. Nature 496:106–109. https://doi.org/10.1038/nature12004
167. Peng T, Hang HC (2015) Bifunctional fatty acid chemical reporter for analyzing S-palmitoylated membrane protein-protein interactions in mammalian cells. J Am Chem Soc 137:556–559. https://doi.org/10.1021/ja502109n
168. Zaro BW, Yang YY, Hang HC, Pratt MR (2011) Chemical reporters for fluorescent detection and identification of O-GlcNAc-modified proteins reveal glycosylation of the ubiquitin ligase NEDD4–1. Proc Natl Acad Sci U S A 108:8146–8151. https://doi.org/10.1073/pnas.1102458108
169. Westcott NP, Hang HC (2014) Chemical reporters for exploring ADP-ribosylation and AMPylation at the host-pathogen interface. Curr Opin Chem Biol 23:56–62. https://doi.org/10.1016/j.cbpa.2014.10.002
170. Grammel M, Luong P, Orth K, Hang HC (2011) A chemical reporter for protein AMPylation. J Am Chem Soc 133:17103–17105. https://doi.org/10.1021/ja205137d
171. Thinon E, Hang HC (2015) Chemical reporters for exploring protein acylation. Biochem Soc Trans 43:253–261. https://doi.org/10.1042/BST20150004

172. Young KH (1998) Yeast two-hybrid: so many interactions, (in) so little time. Biol Reprod 58:302–311
173. Adams PD, Seeholzer S, Ohh M (2002) Identification of associated proteins by coimmunoprecipitation. In: Golemis E (ed) Protein-Protein Interactions. A Molecular Cloning Manual, CSHL Press, New York
174. Li X, Kapoor TM (2010) Approach to profile proteins that recognize post-translationally modified histone "tails". J Am Chem Soc 132:2504–2505. https://doi.org/10.1021/ja909741q
175. Li X, Foley EA, Molloy KR, Li Y, Chait BT, Kapoor TM (2012) Quantitative chemical proteomics approach to identify post-translational modification-mediated protein-protein interactions. J Am Chem Soc 134:1982–1985. https://doi.org/10.1021/ja210528v

Chapter 2
Chemical Reporter for Lysine Malonylation

2.1 Introduction

Lysine malonylation (Kmal), a covalent modification of protein with a malonyl group incorporated at the ε-amine group of lysine through an amide bond (Fig. 2.1), was recently identified as a new PTM based on two different approaches [1, 2]. In 2011, Hao and Lin carried out a careful analysis on the enzymatic activity and structural features of Sirt5, which was previously identified as a weak deacetylase [1]. They found that Sirt5 showed preference for negatively charged carboxylates, like lysine malonylation, rather than uncharged acetyllysine. Several mammalian proteins were identified to have malonylated lysine sites detected by mass spectrometry. In the same year, Zhao et al. performed antibody-based affinity purification of malonylated peptides in conjugation with mass spectrometry further conclusively establish lysine malonylation as a PTM [2]. The modification was then validated by several approaches, including immunoblotting, tandem mass spectrometry and high-performance liquid chromatography and isotope labeling. In consistent with Hao and Lin's discovery, Sirt5 was demonstrated to catalyze demalonylation both in vitro and in vivo [2].

Kmal is chemically different from Kac in two distinct ways. Firstly, the addition of negatively charged malonyl at side chain of lysine can result in change of charge state from $+1$ to -1 under physiological conditions. Interestingly, two well-known protein PTMs, protein phosphorylation and acetylation, can generate a -1 charge on proteins or neutralize the original $+1$ charge, respectively. It is conceivable that lysine malonylation could interfere ionic interactions between the positively charged lysine side chains with negative charged nucleotides, amino acids or other small molecules. In addition, Kmal adds a bulkier moiety to lysine than Kac and is predicted to bring steric hindrance. If located in the same lysine site, lysine malonylation could result in more dramatic effects when compared with lysine acetylation. Therefore, it is necessary to address these possibilities and determine how lysine malonylation exerts its effects on protein functions.

X. Bao, *Study on the Cellular Regulation and Function of Lysine Malonylation, Glutarylation and Crotonylation*, Springer Theses,
https://doi.org/10.1007/978-981-15-2509-4_2

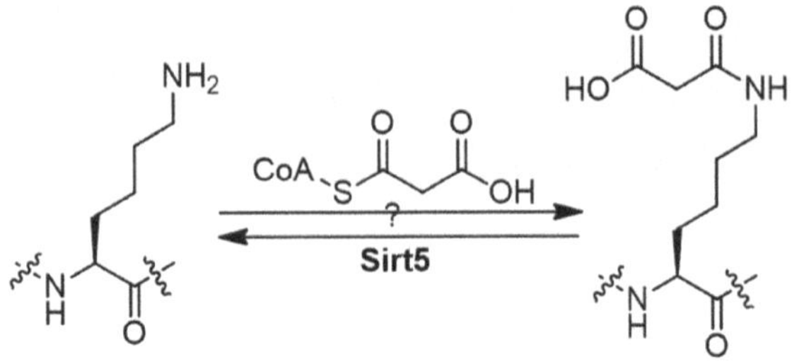

Fig. 2.1 The hypothesized enzymatic reactions for lysine (de)malonylation

A prerequisite to investigate the biological functions of lysine malonylation is to detect and identify malonylated protein substrates. Using a specific antibody, several malonylated proteins, including metabolic enzymes and histones, were identified as lysine malonylated substrates [2]. However, the indirect interactions are always a trouble for antibody-based on assays. The dependence of antibody on surrounding functional groups also restrains the application of antibody for comprehensive profile of targeted proteins. In addition, immunoblotting approaches are not ideal for monitoring malonylation dynamics. To address these difficulties, here we present the development of an alkyne-functionalized chemical reporter for efficient metabolic labeling, robust fluorescent visualization and mass spectrometry-based identification of malonylated proteins.

2.2 Design of the First-Generation Chemical Reporter, Mal-yne, for Kmal

The development of alkyne-carrying chemical reporters has been widely used for substrates profile for a variety of PTMs, such as acetylation [3], lipidation [4–8], glycosylation [9] and AMPylation [10, 11], based on their metabolic incorporation into targeted proteins. There are two features of chemical reporters, one is designed to mimic native metabolites for metabolic labeling onto cellular proteins and the other one is the incorporation of functional groups (i.e. alkyne) for protein visualization and enrichment. For lysine acetylation (Kac), acetate has been reported to be converted into acetyl-CoA, which is then used as donor for lysine acetylation. As a chemical reporter for Kac, the carbon chain backbone in 4-pentynoic acid is designed to serve to as mimic of acetic acid and the alkyne function group is installed for subsequent Cu(I)-catalyzed azide-alkyne cycloaddition. The conjugation on labeled proteins with azide-fluorescent dyes or affinity purification tags enables in-gel visualization and enrichment of targeted proteins prior to identification by mass spectrometry [12] (Fig. 2.2).

Inspired by the development of chemical reporters for the study of protein PTMs and the fact that isotope sodium malonate could be metabolically labeled onto malonylated proteins [2], we developed an alkyne-carrying chemical reporter, 2-propargyl malonate (Mal-yne), based on the structure of malonate for metabolic

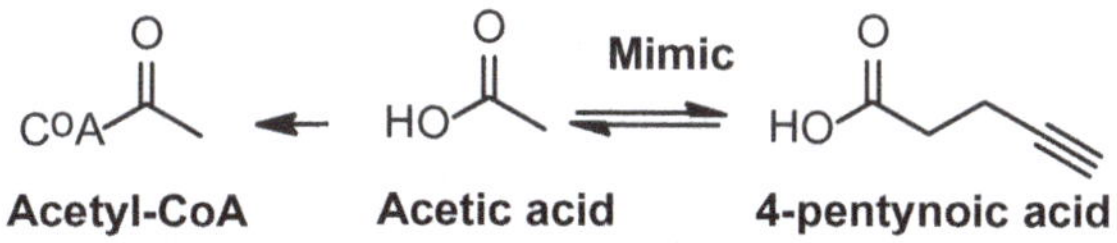

Fig. 2.2 Chemical reporter for profiling acetylated substrates

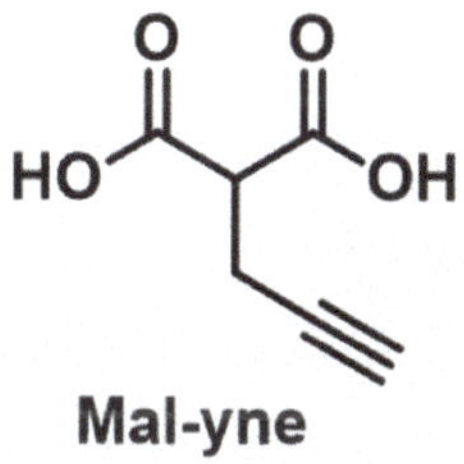

Fig. 2.3 Chemical structure of Mal-yne

labeling of malonylated proteins. In our design, the three-carbon atom backbone aims to mimic malonate and the alkyne handle is used for tag conjugation via 'click' chemistry for fluorescence visualization and identification of malonylated proteins (Fig. 2.3).

2.3 Labeling of Cellular Proteins by Mal-yne

We first examined whether Mal-yne could be metabolic labeled onto cellular proteins (Fig. 2.4). To do this, HeLa S3 cells were treated with 10 mM Mal-yne at indicated time or cultured with a series of concentration of Mal-yne for 6 h. Following metabolic labeling, the whole-cell lysates were extracted and subjected to conjugation of a rhodamine dye to Mal-yne-labeled proteins via an azide-alkyne click chemistry for in-gel fluorescent imaging. As shown in Fig. 2.5, Mal-yne was metabolic labeled onto a diverse spectrum of proteins in both dose- and time-dependent manners. The

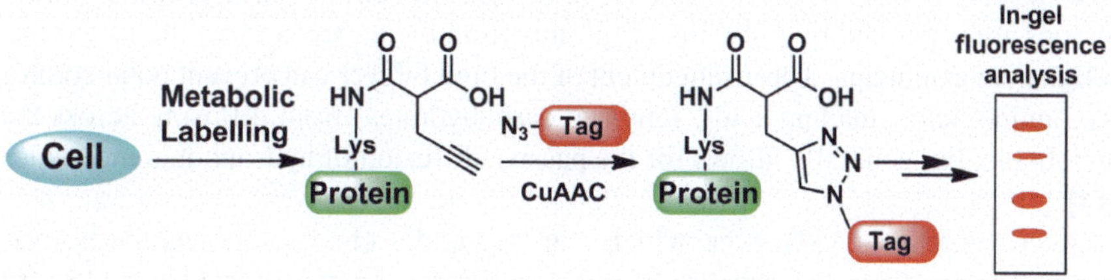

Fig. 2.4 Strategy for detection and identification of malonylated protein substrates using chemical reporter

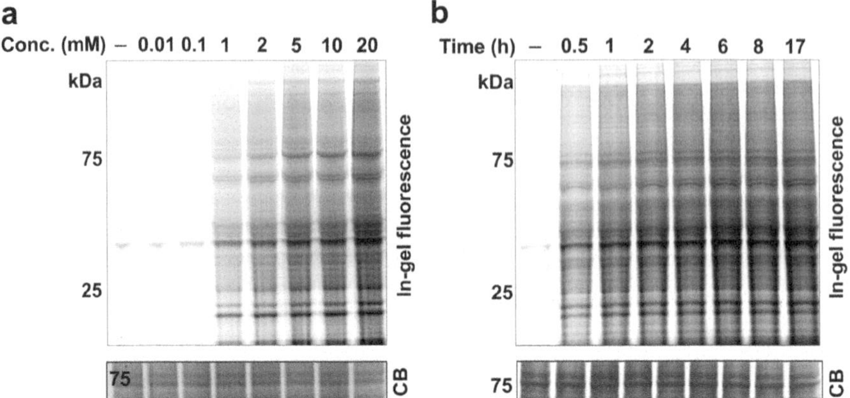

Fig. 2.5 Dose- and time-dependent metabolic labeling of cellular proteins using Mal-yne. **a** Dose-dependent labeling in HeLa S3 cells. HeLa S3 cells were labeled with different concentration of Mal-yne and harvested after 6 h incubation. **b** Time-dependent labeling in HeLa S3 cells. HeLa S3 cells were labeled with 10 mM Mal-yne and harvested after incubation for different time length. PBS was used as negative control. The lysates were reacted with azide-rhodamine and analyzed by in-gel fluorescence scanning. Coomassie blue staining showing the equal loading

optimal labeling condition was revealed to be 10–20 mM of Mal-yne for 4–6 h. The condition is comparable to the metabolic labeling of malonylated proteins using isotopic malonate (20 mM for 24 h)[2].

2.4 Design of the Second-Generation Chemical Reporter, MalAM-yne, for Kmal

We reasoned that the presence of two negative charges of malonate or Mal-yne at physiological pH makes them unfavorable for transport across cell membrane. As a consequence, the relatively high concentration of chemical reporter was needed to achieve optimal metabolic labeling. The cell membrane, as shown in Fig. 2.6, consists primarily of two layers of amphipathic phospholipids [13]. A phospholipid molecule consists of a hydrophilic ‘head’ and two hydrophobic ‘tails’. The hydrophobic ‘tail’ regions are separated by polar ‘head’ groups from the aqueous cytosolic and extracellular environments. The arrangement of the lipid bilayer can prevent polar solutes (i.e. amino acids, nucleic acids, ions and carbohydrates) from diffusing across the membrane, but generally allows for the passive diffusion of hydrophobic molecules [14].

Acetoxymethyl (AM) ester, which can be rapidly cleaved intracellularly, was reported to facilitate the delivery of polar molecules, such as phosphates [15, 16] and MnIIItetra(carboxy-porphyrin) (MnTCP) [17], into cytoplasm. The cleavage of AM esters by intracellular esterases could then release bioactivated molecules,

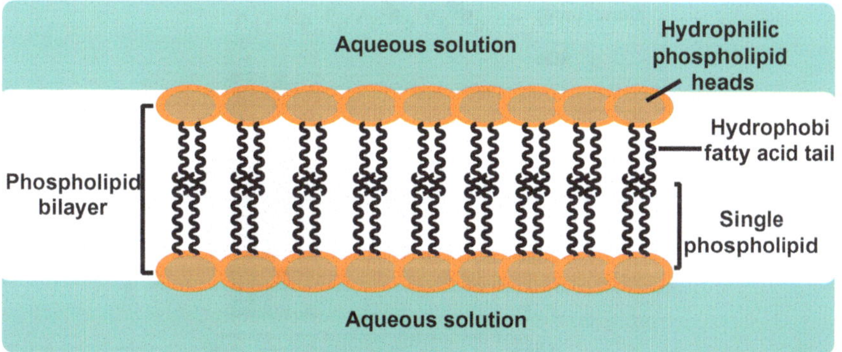

Fig. 2.6 Schematic diagram of a cell membrane

Fig. 2.7 Chemical structure of MalAM-yne

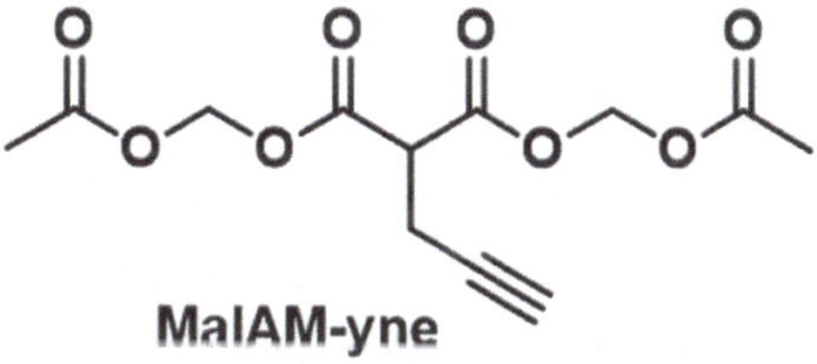

MalAM-yne

cAMP or cGMP and MnTCP. Inspired by this, we designed the second-generation chemical reporter, MalAM-yne (Fig. 2.7), by masking the carboxylates of Mal-yne with two AM groups. We expected that this uncharged new reporter can readily transport across cell membrane and that the subsequent removal of AM groups by cellular esterases would rapidly release bioactive chemical reporter, Mal-yne, for metabolic labeling in cells. To test this hypothesis, HeLa S3 cells were treated with a series of concentration (from 1 μM to 2 mM) of MalAM-yne for only one hour (Fig. 2.8). Following metabolic labeling, the whole-cell lysates were extracted and subjected to click chemistry with azide-rhodamine. After resolution by SDS-PAGE, the fluorescence indicated that hundreds of potential substrates were also labeled by MalAM-yne

2.5 Comparison Between Mal-yne and MalAM-yne

When compared with labeling efficiency of Mal-yne, as we expected, MalAM-yne, at a much lower concentration (50 μM) in as little as 1 h, showed even higher efficiency (Fig. 2.9). More importantly, these two chemical reporters are metabolically equivalent, since the similar labeled protein patterns. We therefore focused on MalAM-yne in later studies.

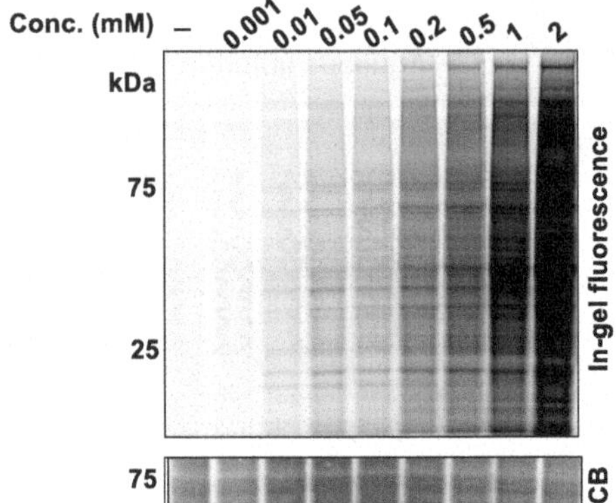

Fig. 2.8 Dose-dependent metabolic labeling on cellular proteins using MalAM-yne. HeLa S3 cells were labeled with different concentration of MalAM-yne and harvested after 1 h incubation. DMSO was used as negative control. The lysates were reacted with azide-rhodamine and analyzed by in-gel fluorescence scanning. Coomassie blue staining showing the equal loading

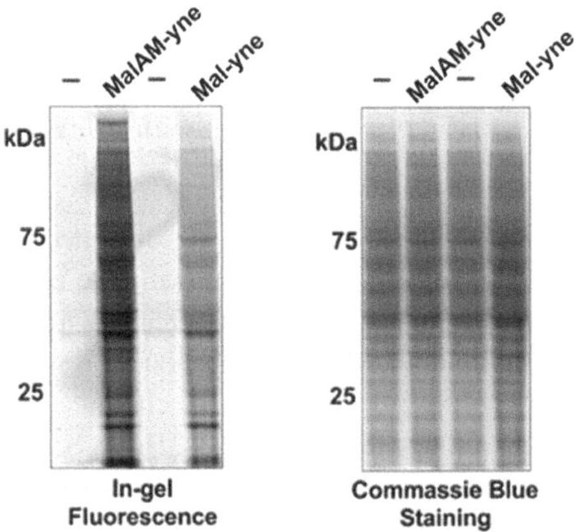

Fig. 2.9 Comparison between chemical reporters Mal-yne and MalAM-yne. HeLa S3 cells were incubated with 20 mM Mal-yne for six hours or 50 μM MalAM-yne for 1 h. DMSO or PBS was used as negative control. The cell lysates were reacted with rhodamine-azide and analyzed by in-gel fluorescent canning. Coomassie blue staining showing the equal loading

2.6 Monitoring Dynamics of Lysine Malonylation

Lysine malonylation, like most PTMs, is a dynamic process. Although it is still unclear of enzymes responsible for transferring malonyl group from malonyl-CoA to targeted residues, Sirt5 was identified as demalonylase, catalyzing the removal of malonyl from lysine [1, 2, 18]. To determine whether MalAM-yne labeling is also reversible, we used a 'pulse-chase' assay. This method has been used for examining a cellular process occurring over time by successively exposing the cells to a labeled

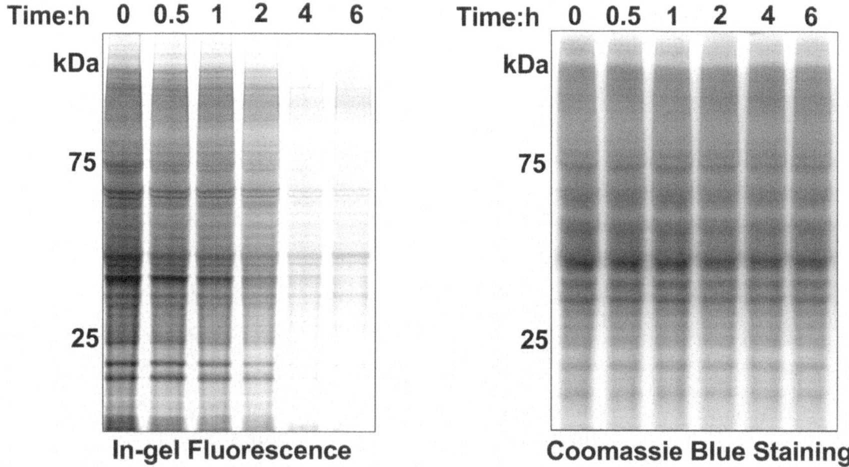

Fig. 2.10 Analysis of dynamics of lysine malonylation using MalAM-yne. HeLa S3 cells were labeled with MalAM-yne for one hour and chased with MalAM for time indicated. The cell lysates were reacted with rhodamine-azide and analyzed by in-gel fluorescent scanning. Coomassie blue staining showing the equal loading

compound (pulse) and then to the same compound in an unlabeled form (chase) [19]. Briefly, HeLa S3 cells were labeled with MalAM-yne for 1 h. After that, fresh medium containing excessive bis(acetoxymethyl) malonate (MalAM) as a competitor was added to remove MalAM-yne. Samples were collected at indicated time points and subjected to cell lysate isolation, click chemistry and in-gel fluorescence scanning. If it was true that those Mal-yne labeled proteins were substrates of lysine malonylation, after introduction of excessive competitor into the environment, the production of protein lysine malonylation would continue, but it would no longer contain Mal-yne introduced in the pulse phase and would not be visible using in-gel fluorescence scanning method. Indeed, as shown in Fig. 2.10, the fluorescence signal of the probe-labeled proteins faded rapidly, indicating the labeling of Mal-yne on proteins is reversible.

2.7 Sirt5 Regulates Lysine Malonylation

We carried out the second experiment to determine whether Sirt5, an identified demalonylase, catalyzes the removal of Mal-yne. To do this, HeLa S3 cells were preincubated with suramin [20], a Sirt5 inhibitor, at concentration of 300 μM for 30 min prior to metabolic labeling using chemical reporter. The samples were collected and underwent similar procedure for fluorescence analysis. The presence of suramin was expected to protect Mal-yne labeling on protein via inhibiting the enzymatic activity of Sirt5. As shown in Fig. 2.11, this treatment resulted in accumulation of MalAM-yne on many proteins.

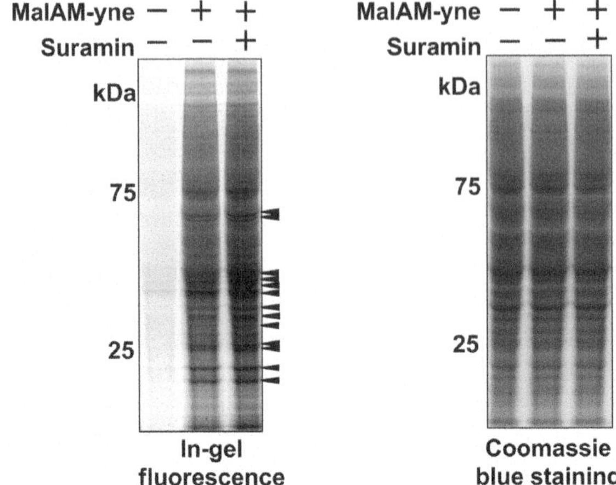

Fig. 2.11 Inhibition of Sirt5 by suramin. HeLa S3 cells were either labeled with 200 μM MalAM-yne alone for 1 h or pre-treated with suramin for 30 min and then labeled with the probe MalAM-yne for another 1 h. Cell lysates were reacted with azide-rhodamine and analyzed by in-gel fluorescence scanning. Those marked proteins showed enhanced labeling intensities. Coomassie blue staining showing the equal loading

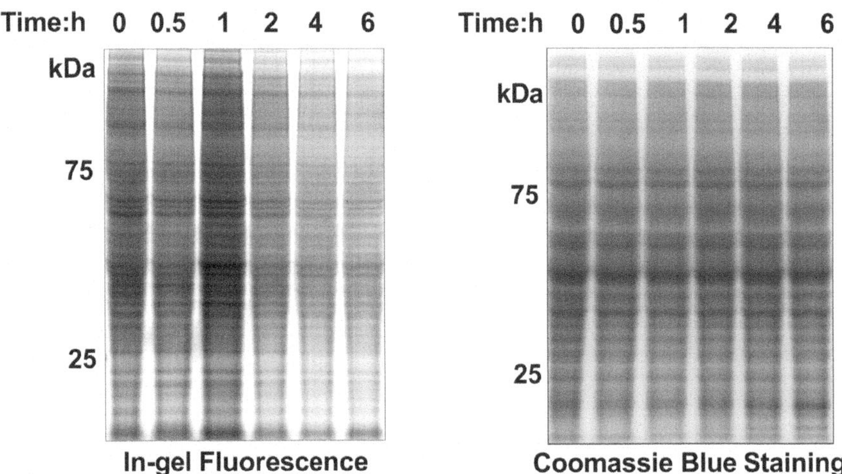

Fig. 2.12 Analysis of the dynamics of lysine malonylation using MalAM-yne. In the presence of a Sirt5 inhibitor, suramin, HeLa S3 cells were labeled with MalAM-yne (200 μM) for one hour and chased with excessive bis(acetoxymethyl) malonate (MalAM) for time indicated. The cell lysates were reacted with rhodamine-azide and analyzed by in-gel fluorescent scanning. Coomassie blue staining showing the equal loading

To further confirm that MalAM-yne-modified proteins are substrates of Sirt5, we combined 'pulse-chase' assay with inhibition of Sirt5. In line with previous results, treatment with suramin significantly retarded the removal of Mal-yne groups from the labeled proteins (Fig. 2.12). These results together demonstrate the fast dynamics of protein lysine malonylation.

2.8 Evolutionary Conservation of Lysine Malonylation

To explore the application scope of our chemical reporter, MalAM-yne, in protein malonylation study, we performed metabolic labeling of MalAM-yne in various cell types derived from different species, including bacteria (*E. coli*), yeast (*S. cerevisiae*), mouse (*Raw 264.7*) and human (HeLa S3, 293T, HCT116, SW480, THP1 and SH-SY5Y). In consistent to previous identification of Kmal in bacteria and human cells [2], the in-gel fluorescent analysis showed robust labeling of a broad spectrum of proteins in *E. coli.* and HeLa S3 cells. Beyond that, MalAM-yne could also be used for metabolic labeling in both yeast and mouse cells, indicating evolutionary conservation of lysine malonylation (Fig. 2.13). Interestingly, the varied protein labeling patterns were present among those cancer cell lines, suggesting a diverse profiles and a dynamic nature of lysine malonylation in mammalian cells and revealing its potential association with diseases (Fig. 2.14).

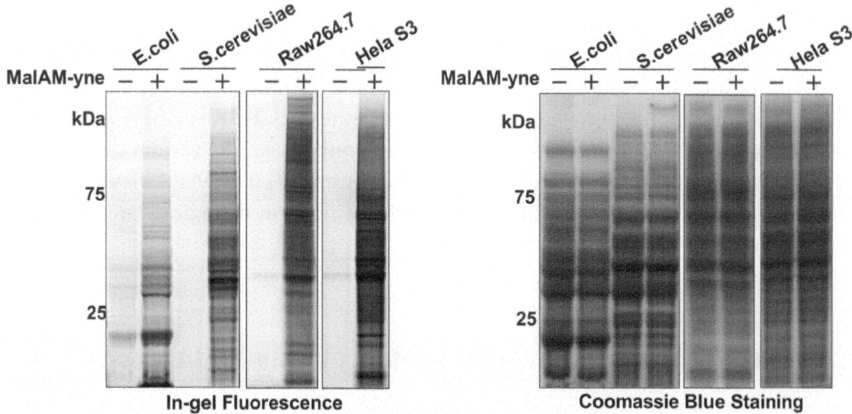

Fig. 2.13 In-gel fluorescence detection of malonylated proteins in cells derived from various species using MalAM-yne. All the different cell lines were labeled with 200 μM MalAM-yne for 1 h, except for *S. cescerevisiae* treated with 1 mM MalAM-yne. DMSO was used as negative control. The lysates were reacted with azide-rhodamine via click chemistry and analyzed by in-gel fluorescence scanning. Coomassie blue staining showing the loading for each lane

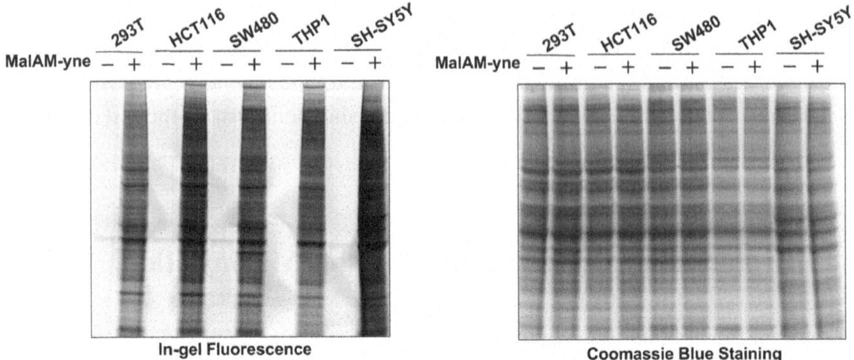

Fig. 2.14 In-gel fluorescence detection of malonylated proteins in different human cell lines using MalAM-yne. All the different cell lines were labeled with 200 μM MalAM-yne for 1 h. DMSO was used as negative control. The lysates were reacted with azide-rhodamine via click chemistry and analyzed by in-gel fluorescence scanning. Coomassie blue staining showing the equal loading

2.9 Immunoblotting Analysis of Known Malonylated Proteins

We then test the ability of MalAM-yne to selectively enrich known malonylation protein substrates. To do this, HeLa S3 were treated with MalAM-yne or DMSO (serving as a negative control). Whole-cell lysate were extracted and conjugated to biotin, an affinity tag, through click chemistry reaction. Taking advantage of high affinity between biotin and streptavidin, those biotin-conjugated proteins were then isolated by streptavidin beads from complex proteomes. Immunoblotting analysis was then carried out on five known malonylated substrates [2], PGK1, HSPA9, HSP90β, GAPDH and NSUN2, using respective specific antibodies. The results showed that all of the five selected proteins could be specifically enriched by MalAM-yne, further confirming the reliability of MalAM-yne in identification of malonylated proteins (Fig. 2.15).

2.10 Profiling of Malonylated Proteins Using MalAM-yne

To identify new malonylated substrates, we performed a proteomics analysis of proteins profiled by MalAM-yne. Simply, after the metabolic labeling, click chemistry and enrichment, eluted proteins were subjected to in-gel trypsin digestion prior to HPLC-MS/MS analysis. The MS/MS data generated from LTQ Orbitrap Velos were submitted to Maxquant for database searching. We found that 14 of all 17 known malonylated proteins [2] were enriched by MalAM-yne in both independent experiments (Table 2.1), including the five proteins confirmed via immunoblotting analysis. In addition, there were another 361 MalAM-yne enriched proteins identified as new candidates of malonylated substrates (Table A.1 in Appendix).

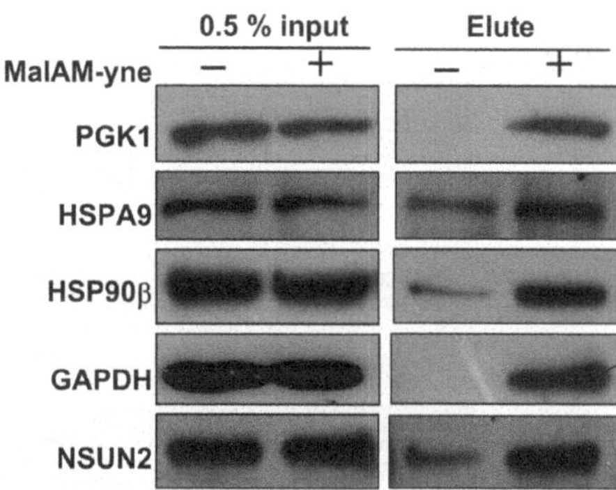

Fig. 2.15 Immunoblotting analysis shows the enrichment of known substrates of lysine malonylation. HeLa S3 cells were incubated with MalAM-yne (200 μM) for one hour. DMSO was used as negative control. The labeled proteins were then conjugated to biotin through click chemistry reaction. Following incubation with streptavidin beads, the enriched proteins were eluted for immunoblotting analysis

Table 2.1 The proteomic analysis for known malonylated proteins enriched by MalAM-yne from two independent experiments

Gene Abbrev.	Expt #1			Expt #2		
	MS/MS count control	MS/MS count probe	MS/MS ratio	Control MS/MS count	Probe MS/MS count	MS/MS ratio
GAPDH	106	178	**1.68**	32	90	**2.81**
SLC25A5	40	121	**3.03**	17	32	**1.88**
NSUN2	9	60	**6.67**	0	14	–
ACTA1	31	44	**1.42**	17	21	**1.24**
PGK1	115	287	**2.50**	92	234	**2.54**
ENO1	332	684	**2.06**	115	267	**2.32**
MDH2	14	50	**3.57**	5	30	**6.00**
GPI	39	117	**3.00**	20	54	**2.70**
HSPA9	16	115	**7.19**	9	70	**7.78**
ALDOA	87	389	**4.47**	47	122	**2.60**
HSPD1	63	254	**4.03**	24	128	**5.33**
HSPE1	0	3	–	0	7	–
HSP90AB1	165	515	**3.12**	81	306	**3.78**
EPRS	66	130	**1.97**	4	61	**15.25**

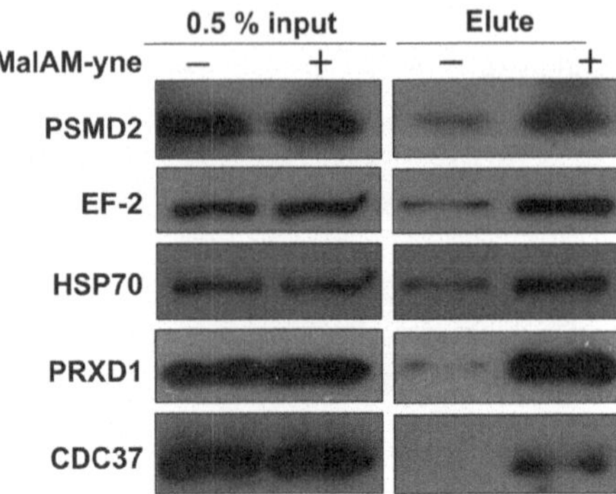

Fig. 2.16 Immunoblotting analysis shows the enrichment of potential substrates of malonylation. HeLa S3 cells were incubated with MalAM-yne (200 μM) for one hour. DMSO was used as negative control. The labeled proteins were then conjugated to biotin through click chemistry. Following incubation with streptavidin beads, the enriched proteins were eluted for immunoblotting analysis

2.11 Immunoblotting Analysis of Candidates of Lysine Malonylation

We also carried out immunoblotting analysis to double confirm the specific enrichment of five candidates in the same manner as mentioned in Sect. 2.9. The enrichment of the five newly identified malonylated protein candidates, including PSMD2, EF2, HSPA1, PRDX1 and CDC37 was also verified as shown in Fig. 2.16.

2.12 Detection of Mal-yne Modified Lysine Sites

To further determine whether MalAM-yne could modify the lysine residues of proteins, we focused on a known malonylated protein, HSP90β that was reported to be malonylated at lysine residue 399 (K399) [2]. To examine the metabolic incorporation of Mal-yne at K399, we did overexpression of streptavidin-tagged HSP90β and subsequent mass spectrometry analysis. Briefly, HEK293T cells with HSP90β overexpression was labeled with MalAM-yne for 1 h. After that, the streptavidin-tagged protein was isolated by Strep-Tactin resin and digested with trypsin. The resulting peptides were then separated and analyzed by tandem HPLC coupled to an LTQ-Orbitrap mass spectrometer. To identify lysine Malyne-modified (KMalyne) peptides, the MS/MS spectra were analyzed by protein sequence alignment. As shown in Fig. 2.17, a peptide (^{393}EMLQQSK MalyneILKVIR^{405}), was identified with a mass shift of 124.0160 Da, which corresponds to the installation of a Mal-yne group at lysine 399. In addition, we also detected satellite peaks with a mass loss of 44 Da for many of the peptide fragments with Malyne-modified lysine. The mass loss corresponds to the loss of CO_2, a reported unique mass signature of

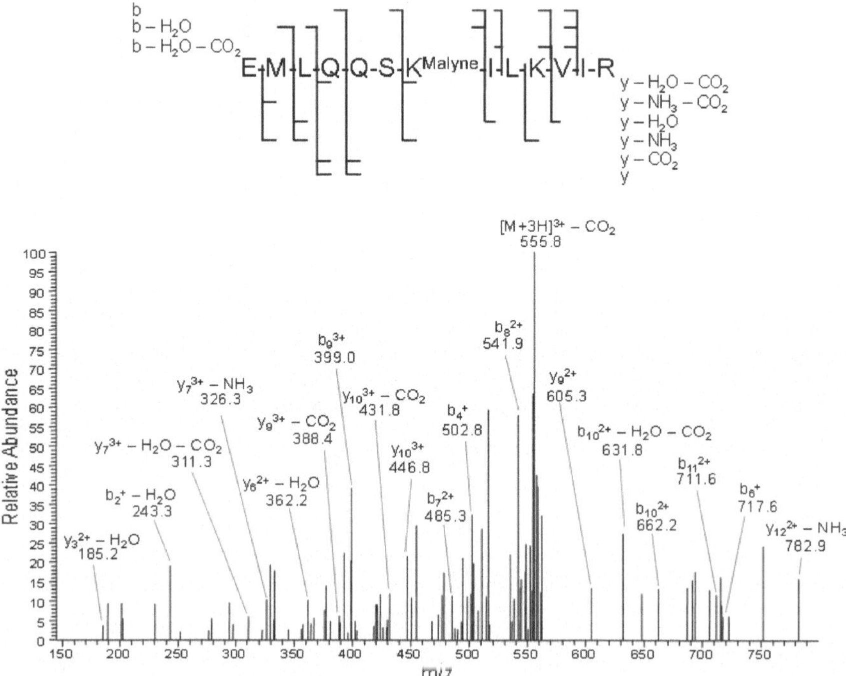

Fig. 2.17 The CID MS/MS spectrum of a triply charged tryptic peptide, [393]EMLQQSK$_{\text{Mal-yne}}$ILKVIR[405], from MalAM-yne modified HSP90 β

lysine-malonylated peptides[2]. This result verified the identification of the KMalyne site and indicated that MalAM-yne can be used for identification of malonylation sites in cells. To identify novel malonylation sites using Mal-yne, we focused on PRDX1, a newly identified malonylated protein candidates. As shown in Fig. 2.18, we identified a peptide with a Malyne-modified lysine, [169]HGEVCPAGWK$_{\text{Malyne}}$ PGSDTIKPDVQK[190], from PRDX1. These experiments together demonstrate the application of MalAM-yne for the detection and identification of lysine malonylation.

2.13 Functional Annotation of Lysine Malonylomes

To understand the biological functions of Kmal proteins, we performed enrichment analysis by using the GO database for Kmal substrates identified in human HeLa S3 cells. The GO biological process analysis of Kmal substrates showed the enrichment in metabolic process (adj $p = 1.106 \times 10^{-13}$), protein translation (adj $p = 2.20 \times 10^{-15}$), protein transport (adj $p = 1.10 \times 10^{-4}$), cell cycle (adj $p = 1.106 \times 10^{-8}$) and response to stress (adj $p = 2.28 \times 10^{-2}$) (Fig. 2.19). The top enriched category for lysine malonylated proteins was metabolic process, which was in consistent with

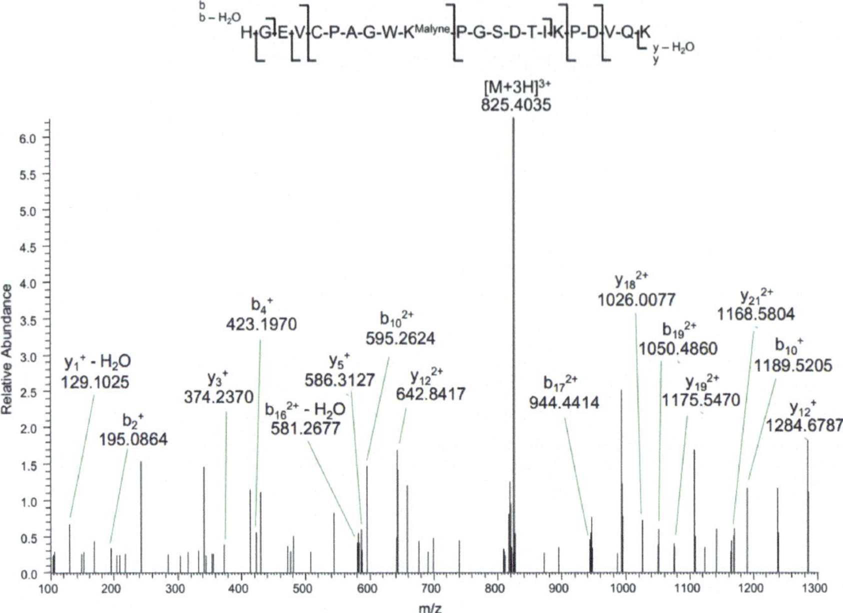

Fig. 2.18 HCD MS/MS spectra of a triply charged tryptic peptide, ^{169}HGEVCPAGWK$_{\text{Malyne}}$PGSDTIKPDVQK^{190}, from MalAM-yne modified PRDX1

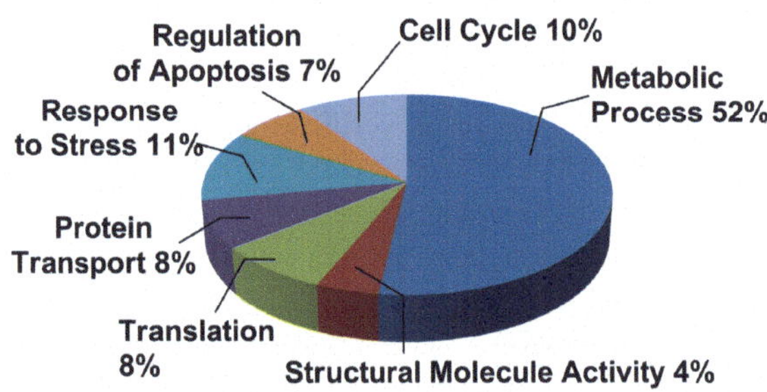

Fig. 2.19 Proteomic analysis of MalAM-yne-metabolically labeled proteins in HeLa S3 cells. Pie charts show the distribution of biological functions of MalAM-yne-labeled proteins

the role of malonyl-CoA as an important metabolic intermediate. The subcellular localization analysis of Kmal substrates showed that Kmal modified proteins were not only localized in mitochondria (9%, adj $p = 3.37 \times 10^{-8}$), but also present in both cytoplasm (adj $p = 2.42 \times 10^{-48}$) and nucleus (adj $p = 9.22 \times 10^{-13}$) with 51% and 30%, respectively (Fig. 2.20). Interesting discovery was that about one-third of malonylated proteins were identified in nucleus, highlighting the potential role of Kmal in modulation of nuclear processes, like RNA binding, telomere maintenance, nuclear division and DNA metabolic process, since proteins involved in those

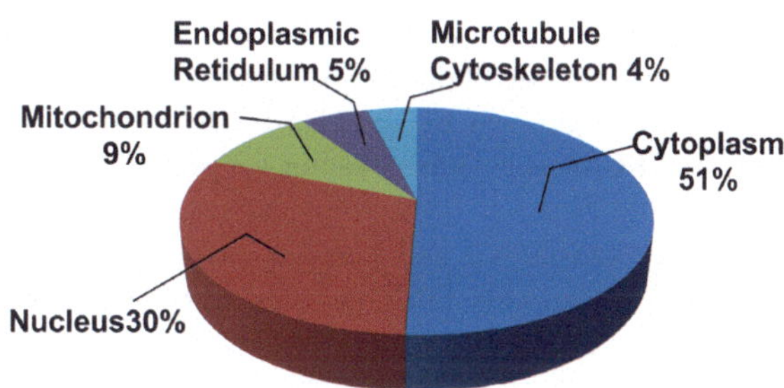

Fig. 2.20 Proteomic analysis of MalAM-yne-metabolically labeled proteins in HeLa S3 cells. Pie charts show the subcellular locations of MalAM-yne-labeled proteins

pathways were identified to be modified by malonylation. This new PTM was also revealed to have potential roles in epigenetic regulations, either serving as a histone mark or a regulator for other histone marks, like Kac, since Kmal was not only identified on histones, but also detected on histone acetyltransferase, deacetylase and histone binding proteins.

2.14 Discussion

In this study, we performed global proteomic analysis of lysine malonylome using chemical reporter, MalAM-yne. Overall, we found 14 of all 17 known malonylated proteins and identified another 361 candidates for lysine malonylation in both independent experiments. One hundred seventy-seven Kmal proteins showed more than five-fold increase and another 51 proteins showed increase between 2 and 5 times. Bioinformatic analysis of malonylated proteins indicated a role in metabolism as their predominant function. Similarly, lysine acetylation has been found in almost every enzyme involved in glucose and fatty acid metabolism. Given that acetylation of metabolic enzymes has been reported to have impact on their enzymatic activity and that Kmal can also occur on metabolic enzymes, lysine malonylation was proposed to also function in regulation of metabolism. Indeed, malonylation on ALDOB in vitro, reported by Wei et al., decreased its enzymatic activity [21]. In addition, elevation of lysine malonylation was observed in type 2 diabetic animal models, including *db/db* mice, *ob/ob* mice, *ob/ob* rat and spontaneously type 2 diabetic monkeys [21].

The identification of a large number of lysine malonylated proteins in cytoplasm and nucleus of human cells suggest the existence of malonyltransferse catalyzing the addition of malonyl group to targeted lysine sites. It has been proposed that non-enzymatic reaction is responsible for posttranslational modifications on mitochondrial proteins, in consideration of the high pH chemical environment and high concentration of metabolic intermediates in mitochondria [22]. However, the pH value in cytosol and nucleus is lower than that of mitochondria, making it undesirable for non-enzymatic malonylation. In addition, Zhao et al. found that subcellular localization of malonylated proteins was very different in liver versus fibroblasts [23]. In the case of MCD +/+ human fibroblasts, malonylated proteins were detected in cytosol and nucleus, while upon MCD −/−, increased localization of malonylated proteins was observed in mitochondria. Another factor, the concentration of malonyl-CoA in cell, which is unknown, makes it hard to determine whether the cellular concentration of malonyl-CoA is a decisive factor for protein malonylation. Therefore, it is highly possible that there exists enzyme-catalyzed lysine malonylation in cell.

Malonyltransferases and demalonylases are required for regulation of lysine malonylation. So far, only Sirt5 has been identified as a demalonylase. However, inhibition of Sirt5 by suramin or knockdown of Sirt5 by siRNA only raised lysine malonylation on several proteins. In addition, Sirt5 was previously reported to exclusively localize in mitochondria, indicating the existence of other unidentified demalonylases.

In summary, we have developed the first chemical reporter, MalAM-yne, for investigation of protein lysine malonylation. Complementary to traditional antibody-based approach, MalAM-yne-based fluorescent detection enables the dynamic analysis of malonylation levels and patterns. MalAM-yne also allows proteomic profiling of malonylated proteins in complex proteomes. Using MalAM-yne, we demonstrate that lysine malonylation is a dynamic, abundant and evolutionarily conserved protein PTM.

2.15 Experimental Methods

2.15.1 Cell Culture

HeLa S3 and 293T cells were cultured in DMEM supplemented with 10% fetal bovine serum (FBS), 100 U/mL penicillin and 100 μg/mL streptomycin. THP1 and SW480 cells were cultured in RPMI medium1640 supplemented with 10% fetal bovine serum (FBS), 100 U/mL penicillin and 100 μg/mL streptomycin. SH-SY5Y cells were cultured in MEM supplemented with 10% fetal bovine serum (FBS), 100 U/mL penicillin and 100 μg/mL streptomycin. HCT116 were cultured in MoCoy's 5A supplemented with 10% fetal bovine serum (FBS), 100 U/mL penicillin and 100 μg/mL streptomycin Cells were maintained in a humidified 37 °C incubator with 5% CO_2.

2.15.2 Metabolic Labeling and Preparation of Cell Lysates

Cells were treated with either Mal-yne or MalAM-yne at concentrations described from 500 mM PBS or 50 μM DMSO stock solutions using the corresponding media for the respective cell types. The same volume of PBS/DMSO was used for the negative control. For co-incubation with inhibitors, HeLa S3 cells were pretreated with suramin (300 μM, 50 mM stock solution in PBS). After 30 min of preincubation at 37 °C, probe MalAM-yne was added to the medium. Following metabolic labeling, cells were washed with ice-cold PBS trice and then lysed with ice-cold modified RIPA lysis buffer (1% NP 40, 150 mM NaCl, 50 mM HEPES, 2 mM $MgCl_2$, 10% Glycerol pH 7.4, EDTA-free Roche protease inhibitor cocktail, 1 mM phenylmethylsulfonylfluoride (PMSF), 0.2U Benzonase). Cell lysates were collected after centrifuging at 16,100 g for 15 min at 4 °C to remove cell debris. Protein concentration was determined by the BCA assay (Navogen). Cell lysates were diluted with IP buffer (150 mM NaCl, 50 mM HEPES, 2 mM $MgCl_2$, 10% Glycerol pH 7.4, EDTA-free Roche protease inhibitor cocktail) to achieve final protein concentration of 1.5 mg/mL for labeling reactions.

2.15.3 Cu(I)-Catalyzed Cycloaddition/Click Chemistry

Briefly, to the prepared samples, 100 μM of rhodamine-azide for in-gel fluorescence scanning or cleavable biotin-azide for streptavidin enrichment were added, followed by 1 mM tris(2-carboxyethyl)phosphine and 100 μM tris[(1-benzyl-1H-1,2,3-triazol-4-yl)methyl]amine, and the reactions were initiated by the addition of 1 mM $CuSO_4$. The reactions were incubated for 1.5 h at room temperature.

2.15.4 In-gel Fluorescence Visualization

The reactions were terminated by the addition of ice-cold acetone (4 volumes), placed at -20 °C for overnight and centrifuged at 6000×g for 5 min at 4 °C to precipitate proteins. The supernatant from the samples was discarded. The protein pellets were washed with ice-cold methanol once, air-dried for ~5 min, resuspended in 20 μL of loading buffer with 50 mM Dithiothreitol (DTT), heated at 85 °C for 8 min, and resolved by SDS-PAGE. The labeled proteins were visualized by scanning the gel on a Typhoon 9410 variable mode imager (excitation 532 nm, emission 580 nm).

2.15.5 Streptavidin Affinity Enrichment of Biotinylated Proteins

After the click chemistry with cleavable biotin-azide, the reaction was quenched by adding 4 volumes of ice-cold acetone to precipitate the proteins. After washing with ice-cold methanol twice, the air-dried protein pellet was dissolved in PBS with 4% SDS, 20 mM EDTA, and 10% glycerol by vortexing and heating. The solution was then diluted with PBS to give a final concentration of SDS of 0.5%. High capacity streptavidin agarose beads (Thermo Fisher Scientific) were added to bind the biotinylated proteins with rotating for 1.5 h at room temperature. To remove non-specific binding, the beads were washed with PBS with 0.2% SDS, 6 M Urea in PBS with 0.1% SDS, and 250 mM NH_4HCO_3 with 0.05% SDS. The enriched proteins were then eluted by incubating with 25 mM $Na_2S_2O_4$, 250 mM NH_4HCO_3, and 0.05% SDS for 1 h. The eluted proteins were dried down with SpeedVac.

2.15.6 Sample Preparation for Mass Spectrometry

The dried proteins were resuspended in 30 μL of lithium dodecyl sulfate sample loading buffer (Life Technologies) with 50 mM dithiothreitol (DTT), heated at 75 °C for 8 min, and then reacted with iodoacetamide in the dark for 30 min to alkylate all of the reduced cysteines. Proteins were then separated on a Bis-Tris gel, followed by fixation in a 50% methanol/7% acetic acid solution. The gel was stained by GelCode Blue stain (Pierce). The diced 1 mm (Goldberg et al. 2007) cubes of gels were then destained by incubating with 50 mM ammonium bicarbonate/50% acetonitrile for 1 h. The destained gel cubes were dehydrated in acetonitrile for 10 min and rehydrated in 25 mM NH_4HCO_3 with trypsin for protein digestion at 37 °C overnight. The resulting peptides were enriched with StageTips. The peptides eluted from the StageTips were dried down by SpeedVac and then resuspended in 0.5% acetic acid for analysis by LC-MS/MS.

2.15.7 Mass Spectrometry

Mass spectrometry was performed on an LTQ-Orbitrap Velos mass spectrometer (Thermo Fisher Scientific). First, peptide samples in 0.1% formic acid were pressure loaded onto a self-packed PicoTip column (New Objective) (360-μm o.d., 75-μm i.d., 15-μm tip), packed with 8–10 cm of reverse-phase C_{18} material (ODS-A C_{18} 5-μm beads from YMC), rinsed for 5 min with 0.1% formic acid and subsequently gradient eluted with a linear gradient from 2 to 35% B in 150 min (A = 0.1% formic acid, B = 100% acetonitrile in 0.1% formic acid, flow rate ~200 nL/min) into the mass spectrometer. The instrument was operated in a data dependent mode cycling through a full scan (300–2000 m/z, single μscan) followed by 10 CID MS/MS scans on the 10 most abundant ions from the immediate preceding full scan. The cations were isolated with a 2-Da mass window and set on a dynamic exclusion list for 60 s after they were first selected for MS/MS. The raw data were processed and analyzed using MaxQuant (version 1.2.2.5). A human fasta file (ipi.HUMAN.v.3.68.fasta) was used as protein sequence searching database. Default parameters were adapted for the protein identification and quantification. In particular, parent peak MS tolerance is 6 ppm, MS/MS tolerance is 0.5 Da, minimum peptide length is 6 amino acids, maximum number of missed cleavages is 2. The proteins identified were supported by at least 2 unique peptides.

2.15.8 Immunoblotting

Proteins separated by SDS-PAGE (10%) were transferred (25 mM Tris, 192 mM Glycine, 20% MeOH in deionized water, 200 mA 2 h) onto a PVDF membrane which was blocked (5% nonfat dried milk and 0.1% Tween-20 in PBS) for 1 h at room temperature or overnight at 4 °C. The membrane was incubated with primary antibody rabbit anti-GAPDH (1:500, Santa Cruz, sc-25778), goat anti-HSP90β (1:500, Santa Cruz, sc-1057), rabbit anti-PGK-1 (1:500, Santa Cruz, sc-17943), rabbit anti-HSPA9 (1:500, Santa Cruz, sc-13967), goat anti-NSUN2 (1:200, Santa Cruz, 83445), rabbit anti-PSMD2 (1:200 Santa Cruz, sc-68352), mouse anti-EF-2 (1:500, Santa Cruz, sc-166415), mouse anti-HSP70 (1:500, Santa Cruz, sc-24), rabbit anti-PRXD1 (1:2000, abcam, ab41906), rabbit anti-CDC37 (1:500, Santa Cruz, sc-5617) diluted in PBST(0.1% Tween-20 in PBS) with 2% BSA and 0.05% NaN3, respectively, followed by washed with PBST for 5 min trice, incubated with goat anti-rabbit-HRP conjugated secondary antibody (1:10000, abcam, ab6721), rabbit anti-goat-HRP conjugated secondary antibody(1:5000, Santa Cruz, sc-2768) or rabbit anti-mouse-HRP conjugated secondary antibody (1:5000, abcam, ab6728) diluted in PBST for 1 h at room temperature, and then detected with chemiluminescent reagents (Thermo).

References

1. Du J et al (2011) Sirt5 is a NAD-dependent protein lysine demalonylase and desuccinylase. Science 334:806–809. https://doi.org/10.1126/science.1207861
2. Peng C et al (2011) The first identification of lysine malonylation substrates and its regulatory enzyme. Mol Cell Proteomics MCP 10(M111):012658. https://doi.org/10.1074/mcp.M111.012658
3. Thinon E, Hang HC (2015) Chemical reporters for exploring protein acylation. Biochem Soc Trans 43:253–261. https://doi.org/10.1042/BST20150004
4. Peng T, Hang HC (2015) Bifunctional fatty acid chemical reporter for analyzing S-palmitoylated membrane protein-protein interactions in mammalian cells. J Am Chem Soc 137:556–559. https://doi.org/10.1021/ja502109n
5. Hang HC, Wilson JP, Charron G (2011) Bioorthogonal chemical reporters for analyzing protein lipidation and lipid trafficking. Acc Chem Res 44:699–708. https://doi.org/10.1021/ar200063v
6. Charron G, Tsou LK, Maguire W, Yount JS, Hang HC (2011) Alkynyl-farnesol reporters for detection of protein S-prenylation in cells. Mol BioSyst 7:67–73. https://doi.org/10.1039/c0mb00183j
7. DeGraw AJ, Palsuledesai C, Ochocki JD, Dozier JK, Lenevich S, Rashidian M, Distefano MD (2010) Evaluation of alkyne-modified isoprenoids as chemical reporters of protein prenylation. Chem Biol Drug Des 76:460–471. https://doi.org/10.1111/j.1747-0285.2010.01037.x
8. Charron G, Zhang MM, Yount JS, Wilson J, Raghavan AS, Shamir E, Hang HC (2009) Robust fluorescent detection of protein fatty-acylation with chemical reporters. J Am Chem Soc 131:4967–4975. https://doi.org/10.1021/ja810122f
9. Zaro BW, Yang YY, Hang HC, Pratt MR (2011) Chemical reporters for fluorescent detection and identification of O-GlcNAc-modified proteins reveal glycosylation of the ubiquitin ligase NEDD4-1. Proc Natl Acad Sci U S A 108:8146–8151. https://doi.org/10.1073/pnas.1102458108

10. Westcott NP, Hang HC (2014) Chemical reporters for exploring ADP-ribosylation and AMPylation at the host-pathogen interface. Curr Opin Chem Biol 23:56–62. https://doi.org/10.1016/j.cbpa.2014.10.002
11. Grammel M, Luong P, Orth K, Hang HC (2011) A chemical reporter for protein AMPylation. J Am Chem Soc 133:17103–17105. https://doi.org/10.1021/ja205137d
12. Yang YY, Ascano JM, Hang HC (2010) Bioorthogonal chemical reporters for monitoring protein acetylation. J Am Chem Soc 132:3640–3641. https://doi.org/10.1021/ja908871t
13. Nagle JF, Tristram-Nagle S (2000) Structure of lipid bilayers. Biochim Biophys Acta 1469:159–195
14. Marsh D (2001) Polarity and permeation profiles in lipid membranes. Proc Natl Acad Sci U S A 98:7777–7782. https://doi.org/10.1073/pnas.131023798
15. Schultz C, Vajanaphanich M, Harootunian AT, Sammak PJ, Barrett KE, Tsien RY (1993) Acetoxymethyl esters of phosphates, enhancement of the permeability and potency of cAMP J Biol Chem 268:6316–6322
16. Schwede F, Brustugun OT, Zorn-Kruppa M, Doskeland SO, Jastorff B (2000) Membrane-permeant, bioactivatable analogues of cGMP as inducers of cell death in IPC-81 leukemia cells. Bioorg Med Chem Lett 10:571–573
17. Haedicke IE et al (2016) An enzyme-activatable and cell-permeable Mn(III)-porphyrin as a highly efficient T1 MRI contrast agent for cell labeling. Chem Sci 7:4308–4317. https://doi.org/10.1039/c5sc04252f
18. Fritzsche S, Springer S (2014) Pulse-chase analysis for studying protein synthesis and maturation. Curr Protoc Protein Sci 78:30–33 (editorial board, John E Coligan [et al]). https://doi.org/10.1002/0471140864.ps3003s78
19. Nishida Y et al (2015) SIRT5 regulates both cytosolic and mitochondrial protein malonylation with glycolysis as a major target. Mol Cell 59:321–332. https://doi.org/10.1016/j.molcel.2015.05.022
20. Schuetz A et al (2007) Structural basis of inhibition of the human NAD+-dependent deacetylase SIRT5 by suramin. Structure 15:377–389. https://doi.org/10.1016/j.str.2007.02.002
21. Du Y et al (2015) Lysine malonylation is elevated in type 2 diabetic mouse models and enriched in metabolic associated proteins. Mol Cell Proteomics MCP 14:227–236. https://doi.org/10.1074/mcp.m114.041947
22. Wagner GR, Payne RM (2013) Widespread and enzyme-independent Nepsilon-acetylation and Nepsilon-succinylation of proteins in the chemical conditions of the mitochondrial matrix. J Biol Chem 288:29036–29045. https://doi.org/10.1074/jbc.M113.486753
23. Colak G et al (2015) Proteomic and biochemical studies of lysine malonylation suggest its malonic aciduria-associated regulatory role in mitochondrial function and fatty acid oxidation. Mol Cell Proteomics MCP 14:3056–3071. https://doi.org/10.1074/mcp.M115.048850

Chapter 3
Identification of Histone Lysine Glutarylation

3.1 Introduction

Cell signaling networks that are involved in regulation of almost all cellular functions are regulated by protein posttranslational modifications (PTMs) [1]. Those covalent chemical modifications allow cells to rapidly carry out actions in response to both external and internal changes. Owing to the rapid development of mass spectroscopy technology, numerous PTMs have been identified [2, 3], including acetylation, methylation, butyrylation, propionylation, phosphorylation, ubiquitination, sumoylation and recently identified crotonylation [4], succinylation [5–7] and malonylation [7, 8]. However, only a few PTMs have been extensively studied. Lysine acetylation (Kac), one of well-studied PTMs, is a conserved and global protein PTM that links cellular metabolism to cell signaling [1, 9].

Lysine glutarylation (Kglu) is newly identified as a protein PTM [10]. Like malonyl-CoA and succinyl-CoA, which are reported as cofactors for lysine malonylation (Kmal) and succinylation (Ksucc), respectively, glutaryl-CoA serves as a donor for lysine glutarylation (Fig. 3.1). However, Kglu is likely to involve in regulation of cellular physiology different from that of Kmal and Ksucc. Malonyl-CoA plays an essential role in the biosynthesis of fatty acid and polyketide. Succinyl-CoA is involved in the TCA cycle and metabolism of odd-numbered fatty acids. Whereas, glutaryl-CoA is an important intermediate in the metabolism of lysine and tryptophan.

The identification of protein substrates of Kglu is crucial to decipher its biological functions. To do this, Zhao et al. performed an antibody-based proteomics approach [10]. By this way, 191 glutarylated proteins were identified in livers from Sirt5 knockout mice. 148 of them were revealed by bioinformatics analysis to locate in mitochondria, coupling lysine glutarylation with cell metabolism. However, there were only 10 glutarylated peptides identified in normal HeLa cell based on this antibody due to its intrinsic limitation. In addition, autoradiography, which was used for visualization of acylated proteins [11], suffers from low sensitivity and is hazardous

X. Bao, *Study on the Cellular Regulation and Function of Lysine Malonylation, Glutarylation and Crotonylation*, Springer Theses,
https://doi.org/10.1007/978-981-15-2509-4_3

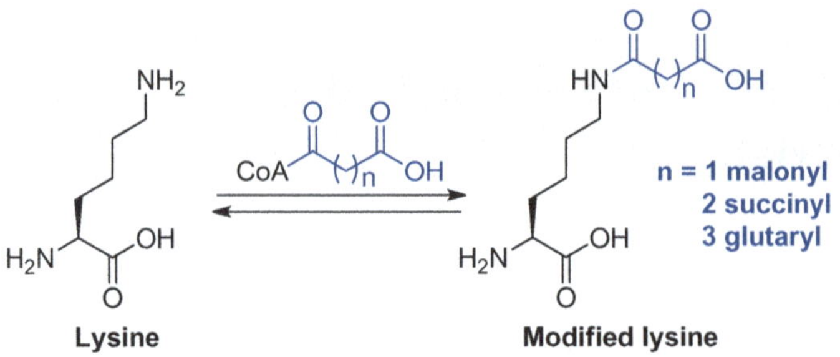

Fig. 3.1 The hypothesis of enzymatic lysine (de)malonylation/succinylation/glutarylation

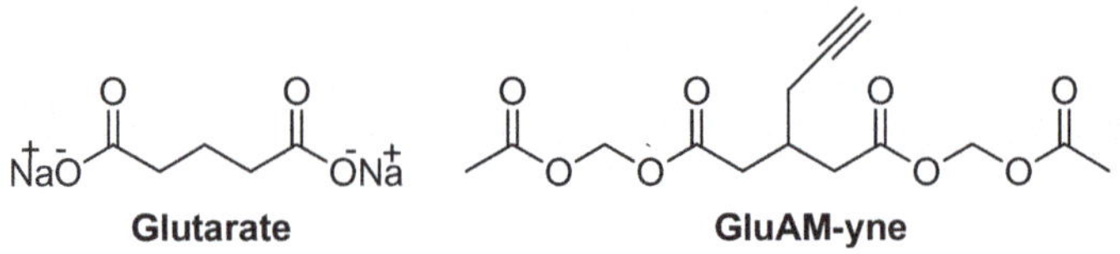

Fig. 3.2 Chemical structures of glutarate and GluAM-yne

to implement. As a complementary approach, azide/alkyne-functionalized chemical reporters have been developed for imaging and proteomics analysis of protein lysine acylation [12, 13]. Here, we developed chemical reporter, GluAM-yne, for analysis of lysine glutarylation and identified as lysine glutarylation as a new histone mark.

3.2 Design of Chemical Reporter, GluAM-yne, for Lysine Glutarylation

Inspired by our previous experience in developing chemical reporter Mal-AM-yne for profiling of malonylated substrates, a chemical reporter, GluAM-yne (Fig. 3.2), was designed for the investigation of lysine glutarylation.

3.3 Metabolic Labeling of Cellular Proteins by GluAM-yne

We firstly examined whether GluAM-yne could be metabolically incorporated into cellular proteins. To do this, dose-dependent labeling efficiency analysis was carried out using HeLa S3 cells. Following metabolic labeling using chemical reporter, the whole-cell lysates were extracted and subjected to azide–alkyne click chemistry to

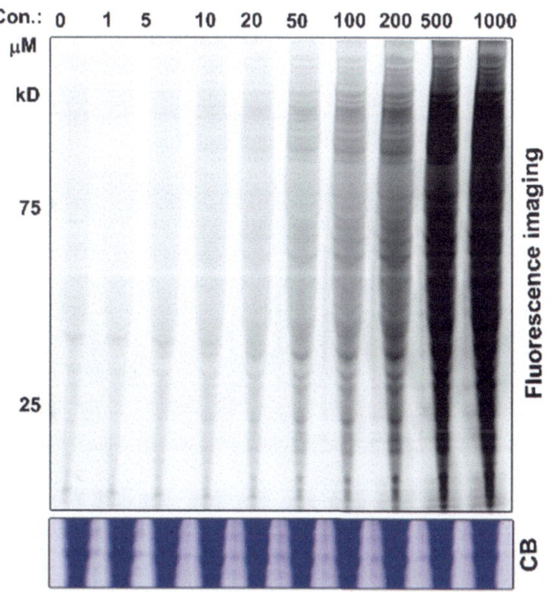

Fig. 3.3 Dose-dependent metabolic labeling of cellular proteins using GluAM-yne. HeLa S3 cells were labeled with different concentration of GluAM-yne and harvested after 1 h incubation. The lysates were reacted with azide-rhodamine and analyzed by in-gel fluorescence scanning. Coomassie blue staining was used as a loading control

conjugate the labeled proteins with fluorescence tag (rhodamine) for in-gel fluorescent imaging analysis. The results revealed that a wide range of proteins was labeled by GluAM-yne at the optimal concentration of 100–200 μM for 1 h (Fig. 3.3).

To validate the feasibility of GluAM-yne in investigation of glutarylated protein substrates, we performed competition assay. Briefly, HeLa S3 cells were treated with 200 μM of GluAM-yne for 1 h with/without preincubation with glutarate. If Glu-yne (a bioactive form after removal of AM esters) could work inside cell in the same way with glutarate, which means that the endogenous enzymes can't differentiate Glu-yne from glutarate, then the added glutarate would compete with GluAM-yne for metabolic labeling onto glutarylated proteins. Indeed, the in-gel fluorescent imaging result revealed that the metabolic labeling of GluAM-yne on proteins could be globally inhibited by co-incubation with glutarate as a competitor (Fig. 3.4).

The ability of GluAM-yne to specifically enrich known modified protein substrates is another indicator of its credibility to profile glutarylated substrates. We therefore selected CSP1 and GAPDH, two known lysine glutarylated proteins, as positive controls. To do this, following the metabolic labeling in HeLa S3 cells using GluAM-yne, the whole-cell lysate were extracted and then subjected to conjugation of a biotin tag via click chemistry. Those labeled proteins were enriched based on the high affinity between biotin and streptavidin. As shown in Fig. 3.5, the immunoblotting analyses showed that both CPS1 and GAPDH could be specifically enriched by our chemical reporter, GluAM-yne. This result together with competition assay demonstrated that GluAM-yne could be used for study of lysine glutarylation.

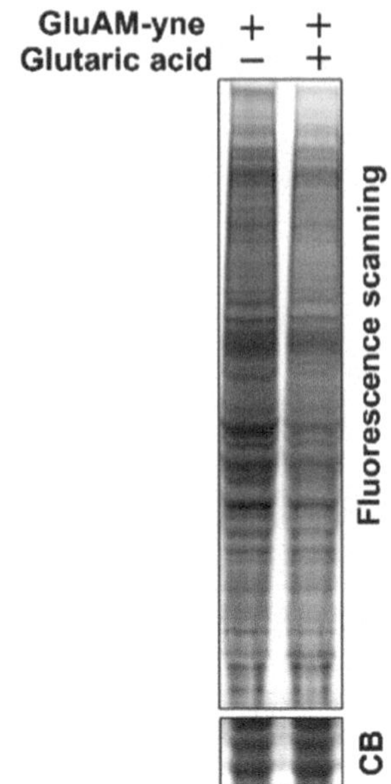

Fig. 3.4 Competition assay using glutaric acid as a competitor. HeLa S3 cells were treated with 200 μM of GluAM-yne for 1 h with/without preincubation with glutarate as a competitor. PBS was used negative control. Extracted whole-cell lysate were then subjected click chemistry and in-gel fluorescence imaging analysis. Coomassie blue staining was used as a loading control

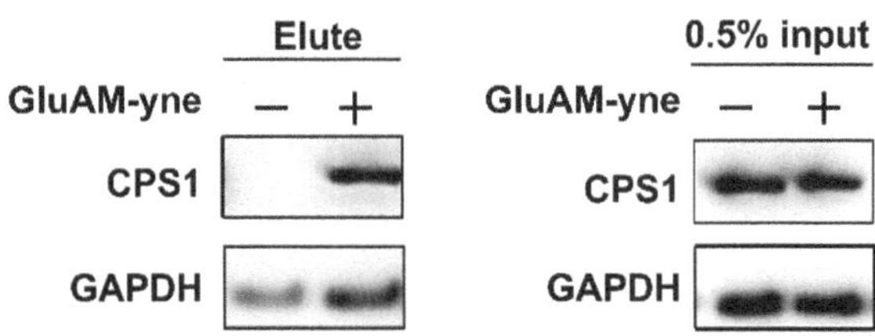

Fig. 3.5 Immunoblotting analysis shows the enrichment of known substrates of glutarylation, CPS1 and GAPDH. HeLa S3 cells were incubated with GluAM-yne (200 μM) for one hour. DMSO was used as negative control. The labeled proteins were then conjugated with a biotin tag through click chemistry. Following incubation with streptavidin beads, the enriched proteins were eluted for immunoblotting analysis

Considering the structure similarity between glutarate and malonate, we carried out experiment to test whether endogenous enzymes can differentiate glutarate from malonate and whether GluAM-yne shares the same protein substrates with MalAM-yne [14], a reporter for lysine malonylation. As shown in Fig. 3.6, the metabolic efficiency of GluAM-yne was comparable to that MalAM-yne (200 μM, 1 h). However, the labeled protein patterns were different between GluAM-yne and MalAM-yne. This result was consistent with the hypothesis that these two protein PTMs are

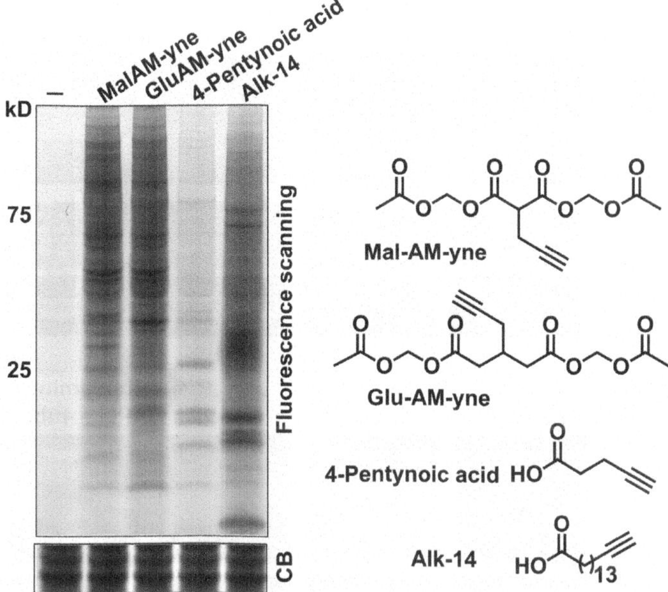

Fig. 3.6 Metabolic labeling efficiency comparison among different chemical reporters, MalAM-yne, GluAM-yne, 4-Pentynoic acid and Alk-14. HeLa S3 cells were incubated with 200 μM GluAM-yne, 200 μM MalAM-yne, 10 mM 4-Pentynoic acid and 20 μM Alk-14 for 1 h (GluAM-yne, MalAM-yne) or 6 h (4-Pentynoic acid, Alk-14). Extracted whole-cell lysate were then subjected to click chemistry and in-gel fluorescence imaging analysis

involved in different signaling pathways, since glutaryl-CoA is an important intermediates in amino acids metabolism while malonyl-CoA plays a key role in fatty acid and polyketide biosynthesis. We also compared the profile of glutarylated proteins with that of other two chemical reporters, 4-Pentynoic acid [12] and Alk-14 [15], which were used for lysine acetylation and long chain fatty acid modification, respectively. The fluorescence imaging analysis showed the profile of labeled proteins by those three chemical reporters were significant different from each other (Fig. 3.6). Collectively, lysine glutarylation is likely to play functional roles distinct for known lysine acylation.

3.4 Metabolic Labeling of Human Histones by GluAM-yne

We then analyzed the subcellular distribution of glutarylated proteins within individual cells. In line with previous report revealing the mitochondrial localization of majority of glutarylated proteins, we also observed the presence of glutarylated proteins in mitochondria (Fig. 3.7). However, the glutarylated proteins were not lim-

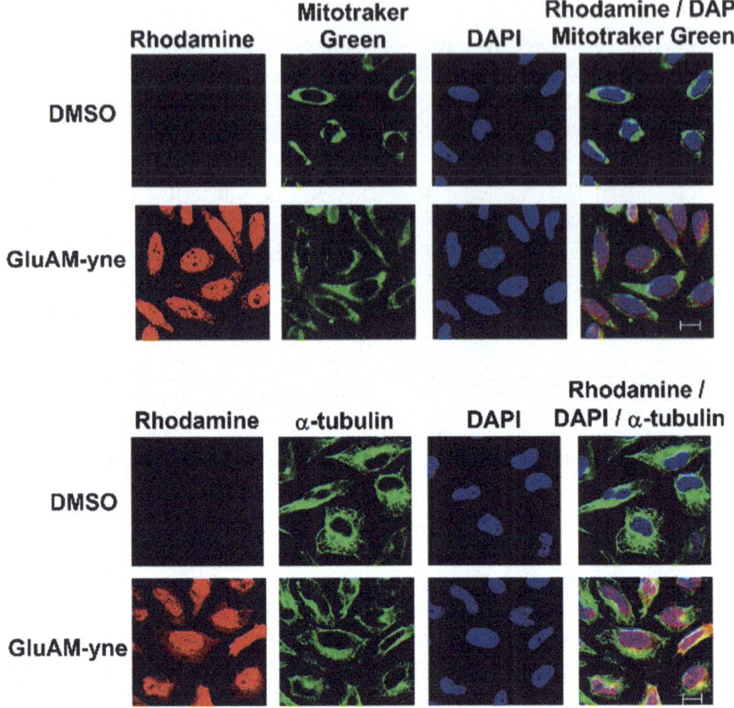

Fig. 3.7 Analysis of GluAM-yne labeled proteins in HeLa cells by fluorescence microscopy. HeLa cells were metabolically labeled with DMSO or 200 μM GluAM-yne for 1 h, fixed, permeabilized and then reacted with 20 μM azide-rhodamine, 1 mM TCEP, 100 μM TBTA and 1 mM $CuSO^4$. Cells were incubated with anti-α-tubulin antibody or mitotraker green overnight at 4 °C. Cells were then subjected to a Zeiss LSM 510 laser scanning confocal microscope. Blue channel: DAPI. Green channel: FITC. Red channel: Rhodamine

ited in mitochondria. Instead, we detected the fluorescence signaling of glutarylated proteins through the whole cell and the heightened signal in nucleus (Fig. 3.7).

In consistent with the result of immunofluorescence analysis, in-gel fluorescence imaging of different cellular fractions also revealed the labeling of GluAM-yne on nuclear proteins (Fig. 3.8). Interestingly, we found there were two protein bands labeled by GluAM-yne with molecular weight similar to that of histones. Considering Kmal and Ksucc have been described as new histone PTMs [16] and the structural similarity of glutaryl with succinyl/malonyl, we speculated that histones could be labeled by GluAM-yne.

We therefore examined whether GluAM-yne could be metabolic labeled onto core histone proteins. To do this, following metabolic labeling in HeLa cells, core histones were extracted and subjected to click chemistry and in-gel fluorescence imaging. As expected, the isolated core histones could be efficiently labeled by GluAM-yne. As shown in Fig. 3.9, there was much stronger fluorescence signal on GluAM-yne treated histones compared with control group treated with only DMSO.

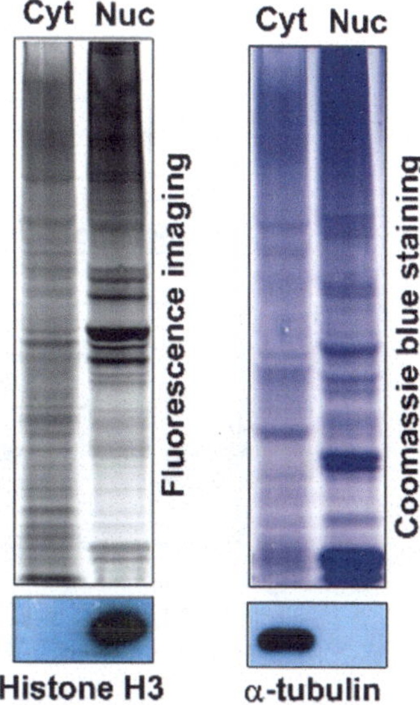

Fig. 3.8 Metabolic labeling of cytosolic or nuclear extraction proteins using GluAM-yne. Immunoblotting analyses of histone H3 and α-tubulin showing the purity of nuclear and cytosolic extraction, respectively

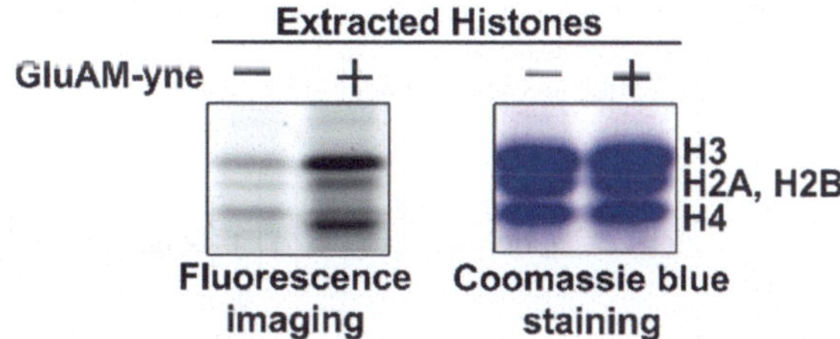

Fig. 3.9 Metabolic labeling of core histone using GluAM-yne. Coomassie blue staining was used as a loading control

3.5 Identification of Histone Lysine Glutarylation

To further map glutarylated lysine sites in histones, we performed mass spectrometry analysis of extracted core histones from HeLa cells. To guarantee high sequence coverage of histone, three parallel methods (Fig. 3.10) were carried out to generate histone proteolytic peptides: (I) the core histones were digested by trypsin in solution without any chemical modifications, (II) following trypsin digestion of core histone, the tryptic peptides underwent in vitro propionylation using propionic anhydride, (III) the core histones were digested by trypsin after in vitro propionylation. Peptides enriched and desalted by stagetips were then separated by HPLC and analyzed with an LTQ-Orbitrap mass spectrometer.

The acquired MS/MS spectra were analyzed by Maxquant software to identify lysine glutarylated peptides. To do this alignment analysis, lysine glutarylation was set as a variable modification. To make sure the reliability of the mass spectrometry results, we did manual inspection of MS/MS spectra for all identified glutarylated histone peptides (see Appendix). As shown in Fig. 3.11, we determined the accurate

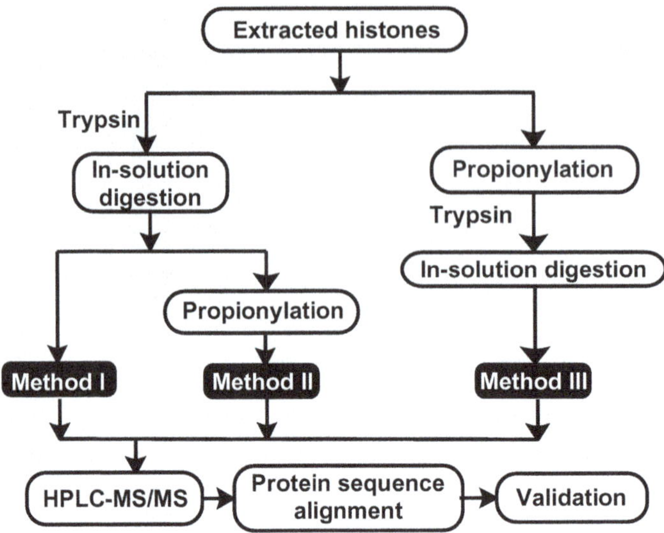

Fig. 3.10 Experimental strategy for identification of histone glutarylation sites. Schematic diagram of the experimental design for comprehensive mapping of PTM sites in core histones from HeLa cells. Histone extracts were in-solution digested with trypsin without any chemical propionylation (Method I), chemically propionylated after (Method II) or before in-solution tryptic digestion (Method III). Peptide mixture were then subjected to mass spectrometry analysis via LTQ-Orbitrap

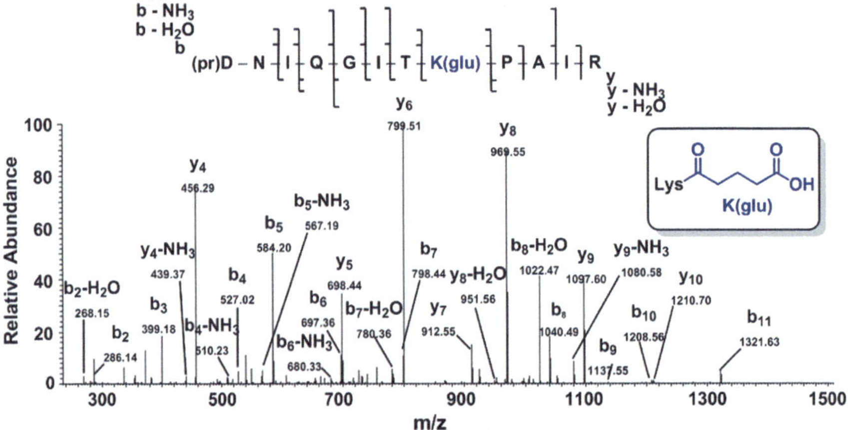

Fig. 3.11 High-resolution MS/MS spectrum of a tryptic peptide, prDNIQGITKgluPAIR, with a mass of +114.0317 Da at its Lys 31 residue identified from in vivo histone H4

glu glu glu glu glu
H2A... EELNKLLGKVTIAQ...AVLLPKKTESHHKAKGK
95 99 118 119 125

glu glu glu glu glu glu glu
H2B...GSKKAVTKAQK...RSRKESYSIYVYKVLKQVH... LAKHAVSEGTKAVTKYTSSK
16 34 43 46 108 116 120

glu glu glu glu glu glu glu glu
H3 ...TGGKAPRKQLATKAARKSAPS...RYQKSTEL...QDFKTDL...IHAKRVTIMPKDIQL...
14 18 23 27 56 79 115 122

glu glu glu glu glu glu glu
H4 ...RGKGGKGLGKGGA...QGITKPAIRR...RGVLKVFL...EHAKRKTVT...VYALKRQG...
5 12 31 59 77 79 91

Fig. 3.12 A diagram showing histone lysine glutarylation sites identified in this study

mass shift was same with predicated glutaryl group. In this experiment, there were 27 glutarylated lysine residues identified on HeLa core histones (Fig. 3.12).

3.6 Validation of Histone Lysine Glutarylation

To verify that the chemical group detected via mass spectrometry represented a PTM induced by lysine glutarylation, we synthesized two standard histone peptides with glutarylated lysine, DNIQGITKgluPAIR, EIAQDFKgluTKLR bearing the same sequence as the in vivo-derived tryptic histone peptides (Fig. 3.13).

Traditionally, endogenous and synthesized peptides are separated by HPLC independently to see if they share the same retention time, which needs a column with high performance to get the exact results from repeated samples [5, 6, 8, 10]. To avoid any unexpected effects from variable column efficiency or manual operation of samples, we decided to carry out in vitro propionylation reaction using two different forms, light ($C_6H_{10}O_3$) and heavy ($C_6D_{10}O_3$), of propionic anhydride. In this way, tryptic peptides from endogenous human histones and synthesized standard peptides were coupled with *N*-terminal propionyl group in light (C_3H_5O) or heavy (C_3D_5O) form, respectively. Those peptides were then combined together for the subsequent MS/MS spectra comparison and HPLC co-elution experiments (Fig. 3.14).

Peptides bearing identical primary sequences and modifications at the same residues should have same MS/MS spectrum and chromatographic properties. The isotopic labeling of a molecule does not change these properties. It means that the labeled molecule has almost the same retention time as its unlabeled counterpart in HPLC experiment and similar signal pattern for MS/MS spectrum, while the mass difference can be differentiated by MS spectrometer. As shown in Fig. 3.15, the mixture of the in vivo and synthetic peptides were co-eluted in HPLC/MS analysis. In the retention time range from 37.69 to 38.68 min, we observed two doubly charged parent ions with mass-to-charge ratios (m/z) of 748.4069 and 750.9224

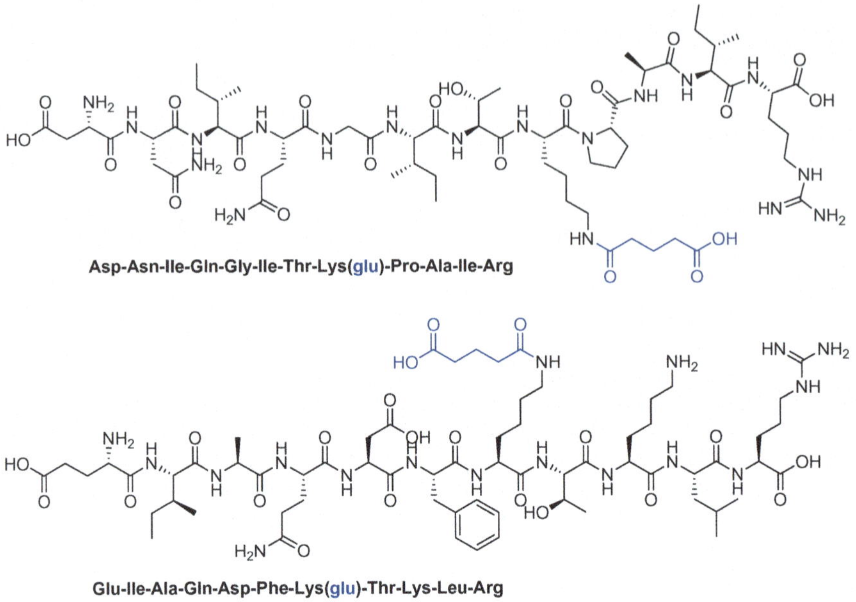

Asp-Asn-Ile-Gln-Gly-Ile-Thr-Lys(glu)-Pro-Ala-Ile-Arg
Glu-Ile-Ala-Gln-Asp-Phe-Lys(glu)-Thr-Lys-Leu-Arg

Fig. 3.13 Chemical structures of synthesized standard peptides, DNIQGITKglu PAIR and EIAQDFKgluTKLR

Tryptic Histone Peptides
In Vitro Propionylation
Synthesized Peptide
Mass Analysis
LC-MS/MS
Time
m/z
5 Da
MS/MS

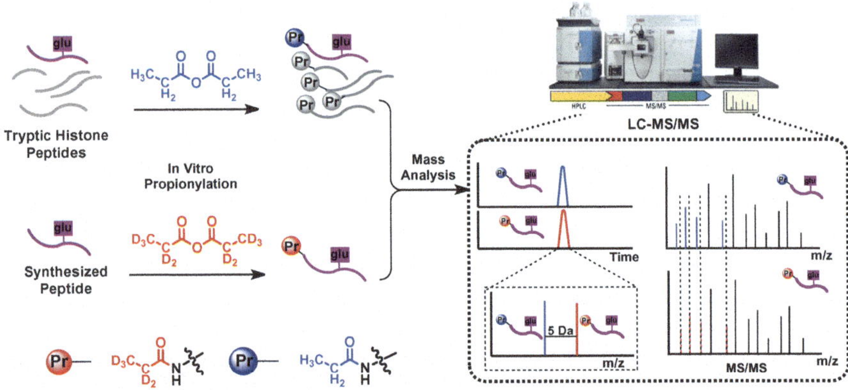

Fig. 3.14 Experimental strategy for validation of histone glutarylation sites. Histone tryptic peptides and synthetic standard peptides were treated with light and heavy forms of propionic anhydride, respectively, and then pooled together for subsequent analysis

that were derived from prDNIQGITKgluPAIR, D_5-prEIAQDFKgluTKLR peptides, respectively. In addition, the in vivo-derived histone peptide exhibited almost identical high-resolution MS/MS spectra with the synthesized peptide with a glutaryl group on lysine residue, expect the difference caused by D_5 labeled propionyl group

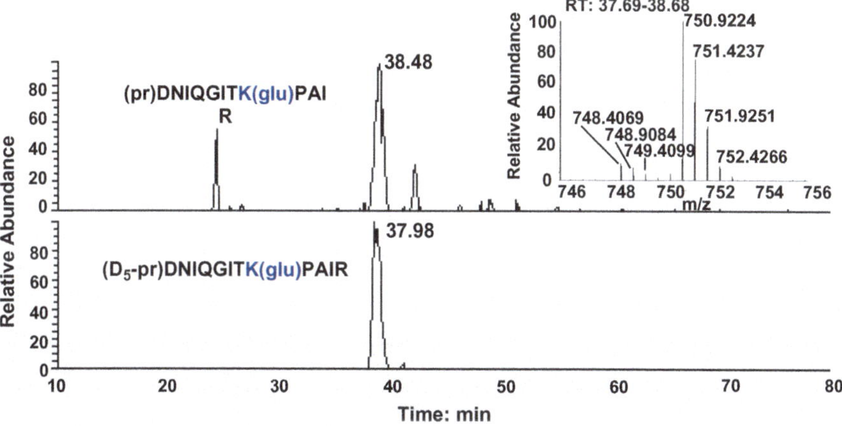

Fig. 3.15 Validation of histone lysine glutarylation. The standard peptide and in vivo-derived peptide showed same retention time

(Fig. 3.17; Tables 3.1 and 3.2). The same analysis was also carried out as a validation of lysine glutarylation at H3 Lys 79 (Figs. 3.16, 3.17 and 3.18; Tables 3.3 and 3.4). These results identified lysine glutarylation as a new histone mark.

3.7 Deglutarylation Activity of Sirt5

Based on the structural similarity between glutarylation and malonylation, Sirt5 was expected to work as a deglutarylase, catalyzing the hydrolysis of glutaryllysine. To test this hypothesis, we synthesized two histone peptides bearing glutarylated or malonylated lysine. Liquid chromatography–mass spectrometry (LC-MS) was used to monitor hydrolysis of the H3K9 modified peptides by Sirt5. As expected, Sirt5 efficiently catalyzed the hydrolysis of both malonylated and glutarylated peptides (Fig. 3.19). This discovery is supported by Zhao's finding that among 18 HDACs, only Sirt5 exhibited significant deglutarylation activity[10].

The identification of lysine glutarylation as a new histone mark and the fact that Sirt5 is, so far, the only identified deglutarylase inspired us to determine whether Sirt5 could serve as a histone deglutarylase. Glutarylation at H3 Lys 9 was not detected on human histones via mass spectrometry analysis, we therefore examined the activity of Sirt5 to hydrolyze a collection of identified glutarylated histone peptides, H2AK95Glu, H2BK120Glu, H3BK116Glu, H3K79Glu, H3K122Glu and H4K31Glu. As shown in Fig. 3.20, Sirt5 manifested varied decrotonylation activities towards these peptides. This substrate selectivity may be partially explained by the sequence-dependent deglutarylation activity of Sirt5.

Table 3.1 Detailed fragmentation table for in vivo derived peptide prDNIQGITKgluPAIR. The theoretical and measured *b*- and *y*-fragment masses for K31 glutarylation of human histone H4 for mass spectrometry shown in Fig. 3.17

m/z	Intensity	Relative intensity	Assignment	Theoretical mass	Error (ppm)
172.0608	1365	8.96	b1	172.061	−1.16
175.1193	1731.7	11.37	y1	175.12	−4
268.0935	1084.4	7.12	− H_2O b2	268.094	−1.87
286.104	10476.1	68.77	b2	286.104	0
399.1881	6099.8	40.04	b3	399.188	0.25
439.2657	837.7	5.5	− NH_3 y4	439.266	−0.68
456.2938	10898.5	71.54	y4	456.293	1.75
527.2488	612.9	4.02	b4	527.247	3.41
567.2363	801.4	5.26	− NH_3 b5	567.241	−8.29
584.2646	5159.6	33.87	b5	584.268	−5.82
697.3511	1829.9	12.01	b6	697.352	−1.29
698.418	3121.6	20.49	y5	698.42	−2.86
799.4661	12670.8	83.18	y6	799.468	−2.38
912.5518	2418.6	15.88	y7	912.552	−0.22
951.5684	766.6	5.03	− H_2O y8	951.563	5.67
969.5729	15233.4	100	y8	969.573	−0.1
1080.6062	2119	13.91	− NH_3 y9	1080.605	1.11
1097.6328	7677.2	50.4	y9	1097.632	0.73

We then examined whether Sirt5 regulates histone lysine glutarylation in cells. To do this, Sirt5 was knockdown by siRNA before metabolic labeling using GluAM-yne. The immunoblotting analysis showed the successful downregulation of Sirt5 (Fig. 3.21a). However, Sirt5 knockdown did not cause an appreciable increase in GluAM-yne labeling on histones (Fig. 3.21b).

3.8 Discussion

The findings presented here identify lysine glutarylation as a new histone modification. Although we have not yet carefully determined the stoichiometry of histone lysine glutarylation, the glutarylated lysine sites were detected by HPLC-MS/MS without any enrichment, indicating its abundance and potential biological significance. Here, we identified 27 glutarylated lysine sites in human histones. Histone lysine glutarylation was validated via comparison the MS/MS fragmentation patterns and HPLC chromatographic profiles between in vivo-derived histone peptides and synthesized peptides bearing glutarylated lysine.

Table 3.2 Detailed fragmentation table for synthesized peptide. The theoretical and measured *b*- and *y*-fragment masses for D_5-prDNIQGITKgluPAIR peptide for mass spectrometry shown in Fig. 3.17

m/z	Intensity	Relative intensity	Assignment	Theoretical mass	Error (ppm)
175.1197	2194.7	5.1	y1	175.1195	1.14
177.0923	2380.9	5.53	b1	177.0922	0.51
291.1356	31183	72.41	b2	291.1351	1.69
404.2201	11104.1	25.78	b3	404.2192	2.20
456.2943	28035	65.1	y4	456.2934	1.97
532.2803	2143.4	4.98	b4	532.2778	4.68
589.2969	12194.1	28.32	b5	589.2992	−3.92
698.4183	7646	17.75	y5	698.4201	−2.58
702.3823	5108.8	11.86	b6	702.3833	−1.44
799.4669	32356.2	75.13	y6	799.4678	−1.13
912.5526	4828.4	11.21	y7	912.5518	0.88
969.5742	43064.9	100	y8	969.5733	0.93
1080.6084	4637.5	10.77	– NH_3 y9	1080.6049	3.24
1097.6343	22591.7	52.46	y9	1097.6319	2.19
1210.7253	2438.4	5.66	y10	1210.7159	7.76
1326.7406	1573.6	3.65	b11	1326.7316	6.78

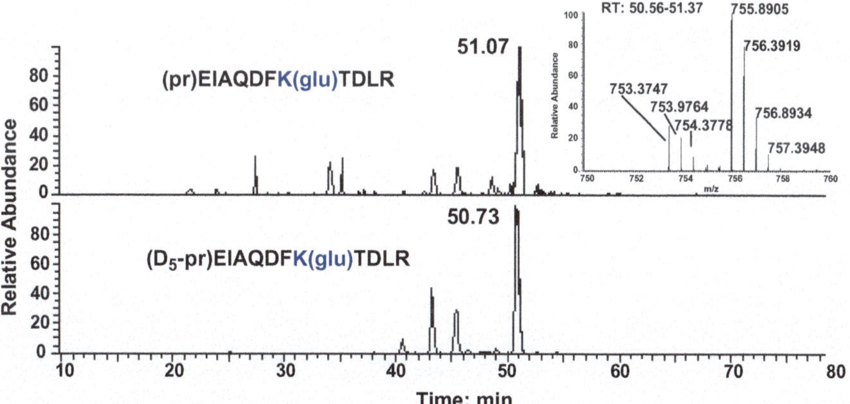

Fig. 3.16 Validation of histone lysine glutarylation. The standard peptide and in vivo-derived peptide showed same retention time

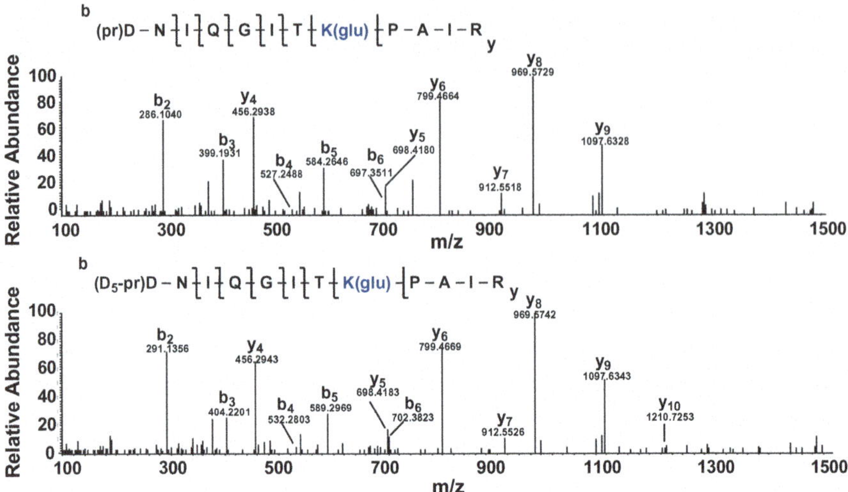

Fig. 3.17 Validation of histone lysine glutarylation. The standard peptide and in vivo-derived peptide show consistent MS/MS spectral pattern, but 5-Da mass difference in their parent peaks

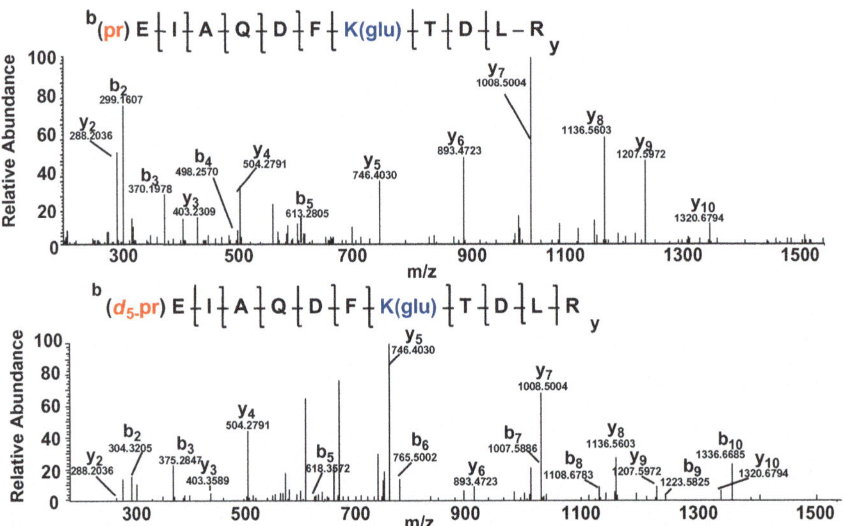

Fig. 3.18 Validation of histone lysine glutarylation. The standard peptide and in vivo-derived peptide show consistent MS/MS spectral pattern, but 5-Da mass difference in their parent peaks

Table 3.3 Detailed fragmentation table for in vivo derived peptide prEIAQDFKgluTDLR. The theoretical and measured *b*- and *y*-fragment masses for K79 glutarylation of human histone H3 for mass spectrometry shown in Fig. 3.18

m/z	Intensity	Relative intensity	Assignment	Theoretical mass	Error (ppm)
175.1194	5092.4	15.23	y1	175.1195	−0.57
186.0766	6165.5	18.44	b1	186.0766	0.00
288.2036	16460.4	49.22	y2	288.2036	0.00
299.1607	24649.9	73.71	b2	299.1607	0.00
370.1978	9168.9	27.42	b3	370.1978	0.00
403.2309	4496	13.44	y3	403.2305	0.99
486.269	875.5	2.62	− H_2O y4	486.2682	1.65
498.257	812.6	2.43	b4	498.2564	1.20
504.2791	10580.2	31.64	y4	504.2782	1.78
613.2805	2080.3	6.22	b5	613.2833	−4.57
728.3905	1225.3	3.66	− H_2O y5	728.3948	−5.90
746.403	11287.8	33.76	y5	746.4048	−2.41
875.4642	1430.3	4.28	− H_2O y6	875.4633	1.03
893.4723	15403.9	46.06	y6	893.4733	−1.12
990.4903	2663.8	7.97	− H_2O y7	990.4902	0.10
1008.5004	33440.1	100	y7	1008.5002	0.20
1118.5516	3386.6	10.13	− H_2O y8	1118.5488	2.50
1119.5314	4260.6	12.74	− NH_3 y8	1119.5318	−0.36
1136.5603	19242.8	57.54	y8	1136.5588	1.32
1189.5884	1792.8	5.36	− H_2O y9	1189.5859	2.10
1190.5675	904.8	2.71	− NH_3 y9	1190.5689	−1.18
1200.5503	860.8	2.57	− H_2O b9	1200.543	6.08
1207.5972	15052.3	45.01	y9	1207.5959	1.08
1302.6699	1075.2	3.22	− H_2O y10	1302.67	−0.08
1320.6794	3665.1	10.96	y10	1320.68	−0.45

Mapping the 27 identified glutarylated lysine sites to the crystal structure of the nucleosome core particle showed the distribution of histone glutarylated lysine is different from well-known histone lysine acetylation. Histone lysine acetylation mainly occurs on histone tails serving as signal to recruit or repel PTM-specific binding proteins, but has limited impact on nucleosome formation [17]. In contrast to histone Kac, most of glutarylated lysine sites are located in globular domains of histones, implying their direct effect on chromatin dynamics and structure. For example, histone H4 lysine 91 lies in the interface between H2A/H2B dimer and H3/H4 tetramer. The substitution of Lys 91 to Ala resulted in enhanced DNA damage sensitivity and relaxed chromatin structure [18]. H3 Lys 115 and Lys 122 lies close to the dyad

Table 3.4 Detailed fragmentation table for synthesized peptide. The theoretical and measured *b*- and *y*-fragment masses for D_5-prEIAQDFKgluTDLR peptide for mass spectrometry shown in Fig. 3.18

m/z	Intensity	Relative intensity	Assignment	Theoretical mass	Error (ppm)
175.1191	6872.1	34	y1	175.1195	−2.28
191.1079	5017.2	24.82	b1	191.1079089	−0.05
288.2032	20212.1	100	y2	288.2036	−1.39
304.1919	18488.8	91.47	b2	304.1919089	−0.03
375.2289	6946	34.37	b3	375.2290089	−0.29
403.2301	5401	26.72	y3	403.2305	−0.99
504.2784	9576.1	47.38	y4	504.2782	0.40
618.3121	2184.2	10.81	b5	618.3146089	−4.06
728.3909	1332.4	6.59	− H_2O y5	728.3948	−5.35
746.4021	6837.5	33.83	y5	746.4048	−3.62
765.3837	1111.8	5.5	b6	765.3830089	0.90
893.4715	7184.9	35.55	y6	893.4733	−2.01
990.4916	2794.5	13.83	− NH_3 b7	990.4826089	9.08
1008.4995	17607.3	87.11	y7	1008.5002	−0.69
1118.5488	1609.7	7.96	− H_2O y8	1118.5488	0.00
1119.5314	3328.5	16.47	− NH_3 y8	1119.5318	−0.36
1136.5596	9453.3	46.77	y8	1136.5588	0.70
1207.5958	7783.2	38.51	y9	1207.5959	−0.08
1320.6841	1187	5.87	y10	1320.68	3.10

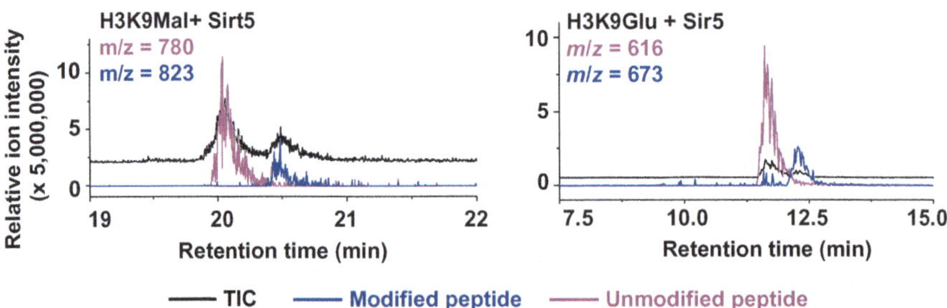

Fig. 3.19 Sirt5 catalyzes the hydrolysis of glutaryl and malonyllysine in vitro. The hydrolysis of the glutarylated and malonylated peptides by Sirt5 was analyzed by liquid chromatography–mass spectrometry. The hydrolysis of H3K9Glu/Mal was observed with Sirt5 in the presence of NAD. Black traces show total ion intensity for all ion species with m/z from 300 to 2000 (i.e., total ion counts, TIC); pink traces show ion intensity (5× magnified) for the masses of unmodified peptides; and blue traces show ion intensity (5× magnified) for the masses of crotonylated peptides

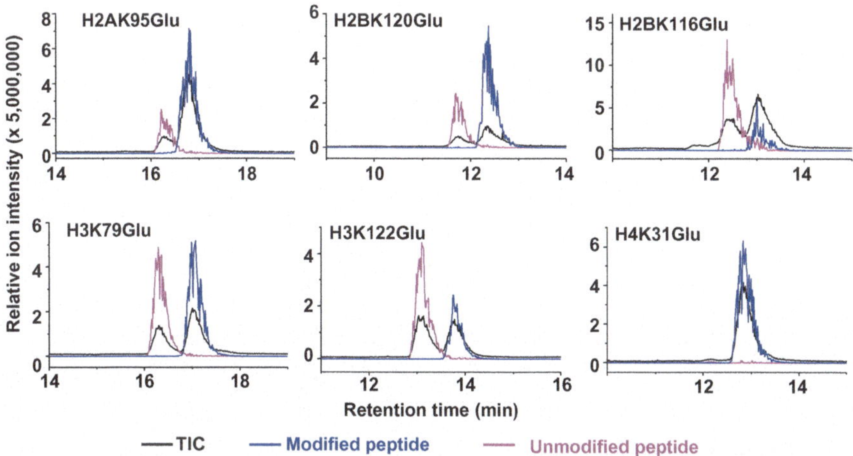

Fig. 3.20 Sirt5 catalyzes the hydrolysis of glutaryl lysine in vitro. Sirt5 showed varied decrotonylation activities towards H2AK95Glu, H2BK120Glu, H3BK116Glu, H3K79Glu, H3K122Glu and H4K31Glu peptides. Black traces show total ion intensity for all ion species with m/z from 300 to 2000 (i.e., total ion counts, TIC); pink traces show ion intensity (5× magnified) for the masses of deglutarylated (unmodified) peptides; and blue traces show ion intensity (5× magnified) for the masses of crotonylated peptides

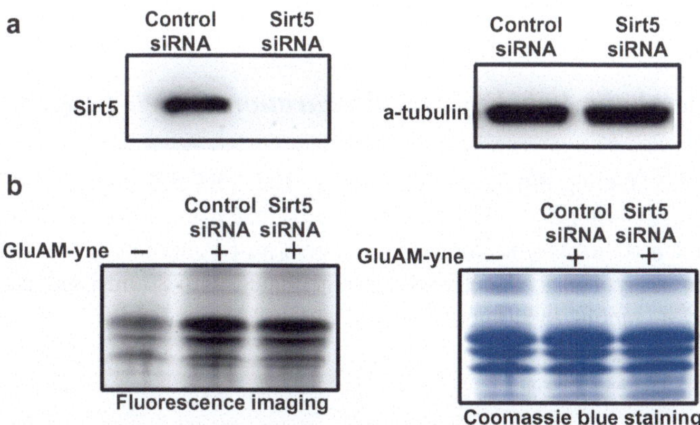

Fig. 3.21 **a** Immunoblotting analysis showing the knockdown of Sirt5. Immunoblotting analysis of α-tubulin was used as loading control. **b** In-gel fluorescence analysis shows the influence of Sirt5 knockdown on GluAM-yne labeling on core histones from HeLa cells. Coomassie blue staining was used as loading control

axis, in which the interaction between histones and DNA is strongest, the modifications on those residues have been reported to reduce DNA–histone octamer affinity and nucleosome stability [19–22]. H3 Lys 56 not only involves in DNA unwrapping process, but also is crucial for histone recognition by histone chaperone complex chromatin assembly factors [23–25].

The reduction or ablation of Sirt5 leads to increases in glutarylation on mitochondrial proteins [10]. On CPS1, eight glutarylated lysine are targeted for removal by Sirt5 and the hyperglutarylation state results in reduced enzymatic activity. However, Sirt5 only regulates 8 Kglu sites out of 44 identified glutarylated lysine sites on CPS1. In this study, using lysine glutarylated histone peptides as substrates, we detected the deglutarylation activity of Sirt5 in vitro. However, fluctuation in Sirt5 expression fails to cause detectable effect on histone glutarylation. Therefore, Sirt5 may only work as one of mitochondrial deglutarylases. Thus, further studies will need to identify nuclear and more mitochondrial deglutarylation enzymes.

3.9 Experimental Methods

3.9.1 Cell Culture

HeLa S3 and HeLa cells were cultured in DMEM supplemented with 10% fetal bovine serum (FBS), 100 U/mL penicillin and 100 μg/mL streptomycin. Cells were maintained in a humidified 37 °C incubator with 5% CO_2.

3.9.2 Metabolic Labeling and Preparation of Cell Lysates

For metabolic labeling efficiency comparison, HeLa S3 cells were incubated with 200 μM GluAM-yne, 200 μM MalAM-yne, 10 mM 4-Pentynoic acid and 20 μM Alk-14 for 1 h (GluAM-yne, MalAM-yne) or 6 h (4-Pentynoic acid, Alk-14). For competition assay, HeLa S3 cells were preincubated with 20 mM glutarate for 6 h followed by metabolic labeling of 200 μM of GluAM-yne for 1 h.

3.9.3 Cu(I)-Catalyzed Cycloaddition/Click Chemistry

Briefly, to the prepared samples, 100 μM rhodamine-azide for in-gel fluorescence scanning or cleavable biotin-azide for streptavidin enrichment was added, followed by 1 mM tris(2-carboxyethyl)phosphine (TCEP), 100 μM tris[(1-benzyl-1H-1,2,3-triazol-4-yl)methyl]amine (TBTA) and finally the reactions were initiated by the addition of 1 mM $CuSO_4$. The reactions were incubated for 1.5 h at room temperature.

3.9.4 In-gel Fluorescence Visualization

The click chemistry reactions were quenched by adding 1 volume of 2× sample buffer. The proteins were heated at 85 °C for 8 min, and resolved by SDS-PAGE. The labeled proteins were visualized by scanning the gel on a Typhoon 9410 variable mode imager (excitation 532 nm, emission 580 nm).

3.9.5 Immunoblotting

Proteins separated by SDS-PAGE were transferred onto a PVDF membrane which was then blocked (5% nonfat dried milk and 0.1% tween-20 in PBS) for 1 h at room temperature. The membrane was incubated with primary antibody, anti-GAPDH (1:500, Santa Cruz), anti-CPS1 (1:500, Santa Cruz) antibodies diluted in PBST (0.1% Tween-20 in PBS) with 2% BSA for overnight at 4 °C, followed by washing with PBST for 5 min trice, incubated with goat anti-rabbit-HRP conjugated secondary antibody (1:20000, Santa Cruz), or rabbit anti-mouse-HRP conjugated secondary antibody (1:5000, Santa Cruz) diluted in PBST for 1 h at room temperature, and then visualized with western blotting detection reagents (Thermo).

3.9.6 Immunofluorescence

HeLa cells grown on cover slips were metabolically labeled with DMSO or 200 μM GluAM-yne for 1 h, fixed with 3.7% PFA, permeabilized with 0.1% Triton X-100 and then reacted with 20 μM azide-rhodamine, 1 mM TCEP, 100 μM TBTA and 1 mM $CuSO_4$. Cells were incubated with anti-α-tubulin antibody or mitotraker green overnight at 4 °C, washed trice with PBST (0.1% tween 20 in PBS) prior FITC-conjugated secondary antibody (containing DAPI for nucleus staining) incubation at room temperature for 1 h. Washed cell were then subjected to a Zeiss LSM 510 laser scanning confocal microscope. Blue channel: DAPI. Green channel: FITC. Red channel: Rhodamine.

3.9.7 Histone Extraction

Acid-extraction method was used to isolate histones from HeLa S3 cells [26]. Briefly, the harvested HeLa S3 cell pellet was resuspended with lysis buffer (10 mM Tris–HCl pH 8.0, 1 mM KCl, 1.5 mM $MgCl_2$ and 1 mM DTT 2 mM PMSF and Roche Complete EDTA-free protease inhibitors) and incubated at 4 °C by rotating for 1 h. The intact nuclei were pelleted by centrifuging at 10,000 g for 10 min at 4 °C. To extract histones, 0.4 N H_2SO_4 was added to resuspend the nuclei, followed by rotating at 4 °C for overnight. After centrifuging to remove the nuclei debris, histones were precipitated by adding 100% trichloroacetic acid (TCA) drop by drop (TCA final concentration: 33%). The precipitated histones were pelleted at 16,000 g for 10 min at 4 °C and washed with ice-cold acetone twice. The air-dried protein pellet was dissolved with ddH_2O and stored at −80 °C for later use.

3.9.8 Sample Preparation for Mass Spectrometry

In-solution tryptic digestion of histone samples was carried out based on previous described protocols. In vitro lysine propionylation of histone extract and tryptic histone peptides were treated with propionic anhydride twice to make sure fully labeling. Histone extracts were in-solution digested without chemical propionylation, chemically propionylated before or after in-solution digestion. The resulting peptides were enriched with the StageTips. The eluted peptides from the StageTips were dried down by SpeedVac and then resuspended in 0.5% acetic acid for analysis by LC-MS/MS.

3.9.9 Mass Spectrometry

Mass spectrometry was performed on an LTQ-Orbitrap Velos mass spectrometer (Thermo Fisher Scientific). First, peptide samples in 0.1% formic acid were pressure loaded onto a self-packed PicoTip column (New Objective) (360-μm o.d., 75-μm i.d., 15-μm tip), packed with 7–10 cm of reverse-phase C18 material (ODS-A C18 5-μm beads from YMC), rinsed for 5 min with 0.1% formic acid and subsequently eluted with a linear gradient from 2 to 35% B in 150 min (A = 0.1% formic acid, B = 0.1% formic acid in ACN, flow rate ~ 200 nL/min) into the mass spectrometer. The instrument was operated in a data dependent mode cycling through a full scan (300–2000 m/z, single μscan) followed by 10 CID MS/MS scans on the 10 most abundant ions from the immediate preceding full scan. The cations were isolated with a 2-Da mass window and set on a dynamic exclusion list for 60 s after they were first selected for MS/MS. The raw data were processed and analyzed using MaxQuant (version 1.2.2.5) setting lysine glutarylation as a variable modification to identify glutarylated lysine sites in histones. A human histone fasta file was used as protein sequence searching database. Default parameters were adapted for the protein identification and quantification. In particular, parent peak MS tolerance is 6 ppm, MS/MS tolerance is 0.5 Da, and minimum peptide length is 6 amino acids, maximum number of missed cleavages is 2. The proteins quantified were supported by at least 2 quantification events.

3.9.10 *Histone Lysine Glutarylation Validation*

In-solution tryptic digestion of histone samples was carried out based on previous described protocols. Tryptic histone peptides and synthesized peptides were treated with light form of propionic anhydride ($C_6H_{10}O_3$) and heavy form of propionic anhydride ($C_6H_{10}O_3$), respectively, and then pooled for peptide enrichment and desalting with the StageTips. The eluted peptides from the StageTips were dried down by SpeedVac and then resuspended in 0.5% acetic acid for analysis by LC-MS/MS.

3.9.11 *Enzymatic Reactions*

The enzymatic activities of Sirt5 were measured by detecting the removal of glutaryl group from peptides. 5 μM of Sirt5 was incubated with 500 μM of corresponding peptides and 1 mM of nicotinamide adenine dinucleotide (NAD) in a reaction buffer containing 20 mM Tris-HCl buffer (pH 7.5) and 1 mM DTT at 37 °C for 2 h. The reactions were stopped by adding 1/3 reaction volume of 20% TFA and frozen in liquid N_2 immediately. Samples were then analyzed by LC-MS with a Vydac 218TP C18 column (4.6 mm × 250 mm, 5 μm, Grace Davison).

References

1. Choudhary C, Weinert BT, Nishida Y, Verdin E, Mann M (2014) The growing landscape of lysine acetylation links metabolism and cell signalling. Nat Rev Mol Cell Biol 15:536–550. https://doi.org/10.1038/nrm3841
2. Walsh C (2006) Posttranslational modification of proteins: expanding nature's inventory. Roberts and Company Publishers, Englewood, CO
3. Walsh CT, Garneau-Tsodikova S, Gatto GJ Jr (2005) Protein posttranslational modifications: the chemistry of proteome diversifications. Angew Chem 44:7342–7372. https://doi.org/10.1002/anie.200501023
4. Tan M et al (2011) Identification of 67 histone marks and histone lysine crotonylation as a new type of histone modification. Cell 146:1016–1028. https://doi.org/10.1016/j.cell.2011.08.008
5. Zhang Z, Tan M, Xie Z, Dai L, Chen Y, Zhao Y (2011) Identification of lysine succinylation as a new post-translational modification. Nat Chem Biol 7:58–63. https://doi.org/10.1038/nchembio.495
6. Colak G et al (2013) Identification of lysine succinylation substrates and the succinylation regulatory enzyme CobB in *Escherichia coli*. Mol Cell Proteomics MCP 12:3509–3520. https://doi.org/10.1074/mcp.M113.031567
7. Du J et al (2011) Sirt5 is a NAD-dependent protein lysine demalonylase and desuccinylase. Science 334:806–809. https://doi.org/10.1126/science.1207861
8. Peng C et al (2011) The first identification of lysine malonylation substrates and its regulatory enzyme. Mol Cell Proteomics MCP 10(M111):012658. https://doi.org/10.1074/mcp.M111.012658
9. Verdin E, Ott M (2015) 50 years of protein acetylation: from gene regulation to epigenetics, metabolism and beyond. Nat Rev Mol Cell Biol 16:258–264. https://doi.org/10.1038/nrm3931

10. Tan M et al (2014) Lysine glutarylation is a protein posttranslational modification regulated by SIRT5. Cell Metab 19:605–617. https://doi.org/10.1016/j.cmet.2014.03.014
11. Brownell JE, Allis CD (1995) An activity gel assay detects a single, catalytically active histone acetyltransferase subunit in Tetrahymena macronuclei. Proc Natl Acad Sci U S A 92:6364–6368
12. Yang YY, Ascano JM, Hang HC (2010) Bioorthogonal chemical reporters for monitoring protein acetylation. J Am Chem Soc 132:3640–3641. https://doi.org/10.1021/ja908871t
13. Thinon E, Hang HC (2015) Chemical reporters for exploring protein acylation. Biochem Soc Trans 43:253–261. https://doi.org/10.1042/BST20150004
14. Bao X, Zhao Q, Yang T, Fung YM, Li XD (2013) A chemical probe for lysine malonylation. Angew Chem 52:4883–4886. https://doi.org/10.1002/anie.201300252
15. Wilson JP, Raghavan AS, Yang YY, Charron G, Hang HC (2011) Proteomic analysis of fatty-acylated proteins in mammalian cells with chemical reporters reveals S-acylation of histone H3 variants. Mol Cell Proteomics MCP 10:M110 001198. https://doi.org/10.1074/mcp.m110.001198
16. Xie Z et al (2012) Lysine succinylation and lysine malonylation in histones. Mol Cell Proteomics MCP 11:100–107. https://doi.org/10.1074/mcp.M111.015875
17. Fidlerova H, Kalinova J, Blechova M, Velek J, Raska I (2009) A new epigenetic marker: the replication-coupled, cell cycle-dependent, dual modification of the histone H4 tail. J Struct Biol 167:76–82. https://doi.org/10.1016/j.jsb.2009.03.015
18. Ye J et al (2005) Histone H4 lysine 91 acetylation a core domain modification associated with chromatin assembly. Mol Cell 18:123–130
19. Masumoto H, Hawke D, Kobayashi R, Verreault A (2005) A role for cell-cycle-regulated histone H3 lysine 56 acetylation in the DNA damage response. Nature 436:294–298. https://doi.org/10.1038/nature03714
20. Ozdemir A, Spicuglia S, Lasonder E, Vermeulen M, Campsteijn C, Stunnenberg HG, Logie C (2005) Characterization of lysine 56 of histone H3 as an acetylation site in Saccharomyces cerevisiae. J Biol Chem 280:25949–25952. https://doi.org/10.1074/jbc.c500181200
21. Zhang L, Eugeni EE, Parthun MR, Freitas MA (2003) Identification of novel histone post-translational modifications by peptide mass fingerprinting. Chromosoma 112:77–86. https://doi.org/10.1007/s00412-003-0244-6
22. Manohar M et al (2009) Acetylation of histone H3 at the nucleosome dyad alters DNA-histone binding. J Biol Chem 284:23312–23321. https://doi.org/10.1074/jbc.m109.003202
23. Recht J et al (2006) Histone chaperone Asf1 is required for histone H3 lysine 56 acetylation, a modification associated with S phase in mitosis and meiosis. Proc Natl Acad Sci U S A 103:6988–6993. https://doi.org/10.1073/pnas.0601676103
24. Han J, Zhou H, Li Z, Xu RM, Zhang Z (2007) Acetylation of lysine 56 of histone H3 catalyzed by RTT109 and regulated by ASF1 is required for replisome integrity. J Biol Chem 282:28587–28596. https://doi.org/10.1074/jbc.m702496200
25. Hiraga S, Botsios S, Donaldson AD (2008) Histone H3 lysine 56 acetylation by Rtt109 is crucial for chromosome positioning. J Cell Biol 183:641–651. https://doi.org/10.1083/jcb.200806065
26. Shechter D, Dormann HL, Allis CD, Hake SB (2007) Extraction, purification and analysis of histones. Nat Protoc 2:1445–1457. https://doi.org/10.1038/nprot.2007.202

Chapter 4
Glutarylation at Histone H4 Lysine 91 Modulates Chromatin Assembly

4.1 Introduction

The genomic information of eukaryotic cells is stored in chromatin, a compacted structure formed by packaging DNA around proteins called histones [1, 2]. The organization of chromatin is not static but highly dynamic [3, 4]. By affecting DNA accessibility, the dynamic structure of chromatin plays essential roles in many fundamental nuclear processes, including gene transcription, DNA replication and DNA damage repair [5, 6]. One primary mechanism to control chromatin dynamics is post-translational modifications (PTMs) on histones [7, 8]. Mounting evidence suggests that histone modifications can modulate distinct steps of the dynamic organization of chromatin, including histone folding, assembly of DNA and histones into nucleosome and compaction of nucleosomes into higher order structures of chromatin [9]. Histone modifications play a crucial role in regulating chromatin-templated biological processes. The aberrant histone modification profiles often lead to defects in gene expression and genome stability, which may give rise to human diseases such as cancer and developmental disorder [10].

There are two mechanisms proposed to elucidate the contribution of histone PTMs in regulation of chromatin dynamics and chromatin-associated processes [11]. One is the direct modulation from histone PTMs on chromatin compaction by altering the histone-DNA and histone-histone interactions, thereby changing chromatin structure and regulating the access of DNA-binding proteins. Alternatively, histone PTMs regulate chromatin structure and function by recruiting or repelling effector proteins to or from chromatin, which are in turn translated into biological outcomes.

Lysine glutarylation (Kglu) is newly identified as a histone PTM. Similar to lysine succinylation (Ksucc) and malonylation (Kmal), which were recently identified on histones [12], the negatively charged carboxylate group in Kglu could not only cause change of charge state from $+1$ to -1, but also provide potential steric hindrance.

X. Bao, *Study on the Cellular Regulation and Function of Lysine Malonylation, Glutarylation and Crotonylation*, Springer Theses,
https://doi.org/10.1007/978-981-15-2509-4_4

The installation of those negatively charged PTMs was expected to be a great challenge to chromatin stability and therefore affect chromatin-associated nuclear processes. Indeed, mutation on histone H3K79 or H4K77 to Glu, a negatively charged amino acid, resulted in a loss of silencing at telomeres and rDNA. Replacement of H2AK21 with Glu, but not Ala, was sensitive to DNA damage reagent. In addition, the substitutions of histone H4K31 with Glu, significantly reduces cell viability [12].

In this study, we focused on glutarylation at histone H4 lysine 91. The significance of positively charged H4 Lys 91, a conserved residue from yeast to human (Fig. 4.1), is highlighted by the fact that this residue can form a salt bridge with a glutamic acid residue in histone H2B to maintain the stability of nucleosome (Fig. 4.2). The replacement of H4K91 with alanine or glutamine that mimic the constitutively acetylated state was reported to significantly increase DNA damage sensitivity and accessibility of nuclease to DNA substrate in chromatin. We reasoned that glutarylation on lysine would have more potential to destabilize nucleosome structure to create a more relaxed environment for chromatin remodelers and promote DNA accessibility.

```
sp|P02309|H4_YEAST  1   MSGRGKGGKGLGKGGAKRHRKILRDNIQGITKPAIRRLARRGGVKRISGL 50
sp|P62805|H4_HUMAN  1   MSGRGKGGKGLGKGGAKRHRKVLRDNIQGITKPAIRRLARRGGVKRISGL 50
                        *********************:****************************

sp|P02309|H4_YEAST  51  IYEEVRAVLKSFLESVIRDSVTYTEHAKRKTVTSLDVVYALKRQGRTLYG 100
sp|P62805|H4_HUMAN  51  IYEETRGVLKVFLENVIRDAVTYTEHAKRKTVTAMDVVYALKRQGRTLYG 100
                        ****.*.*** ***.****:*************::***************

sp|P02309|H4_YEAST  101 FGG 103
sp|P62805|H4_HUMAN  101 FGG 103
                    ***
```

Fig. 4.1 Sequence alignment of histone H4 between yeast and human. Highlighted in red is K91 in H4

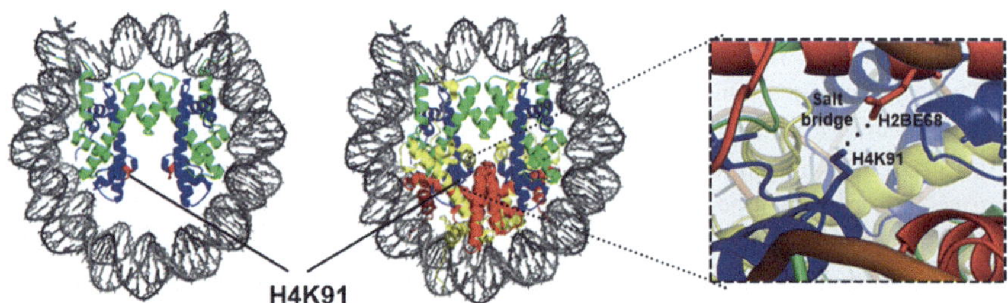

Fig. 4.2 Position of H4K91 in nucleosome (3AFA). K91 is highlighted as a red sphere, localized at the interface between H2A/H2B dimers and H3/H4 tetramer. Histone H4 is in dark blue, H3 in green, H2A in yellow and H2B in red. H4K91 can form a salt bridge with a H2BE68 to maintain the stability of nucleosome

4.2 Generation of Yeast Strains Containing H4K91 Mutants

To gain insight into the potential functions of histone Kglu, we generated budding yeast strains containing mutants from lysine (K) into glutamate (E), glutamine (Q) and arginine (R), respectively (Fig. 4.3). Mutation from lysine to glutamate (H4K91E) was to simulate glutarylated lysine. The replacement of lysine to glutamine (H4K91Q) was expected to as a mimic of acetylated lysine. Positively charged arginine (H4K91R) was a substitution of unmodified lysine.

For the generation of yeast strains with H4K91 alleles, gene deletion and site-specific mutation were done by PCR, utilizing linear DNA fragments that contain homology to the target gene and a selectable marker (KanMX6 or Trp). Considering there are two genes, *HHF1* and *HHF2*, encoding histone H4 in yeast, to construct yeast strain with H4K91E mutant, we did gene deletion of *HHF2* with *Trp* marker, a selective marker for auxotrophic strain, and carried out site-directed mutation in *HHF1* with the aid of *KanMX6*, an antibiotic marker. Similar procedures were carried out to construct other H4K91Q/R strains (Figs. 4.4 and 4.5); (Table 4.1).

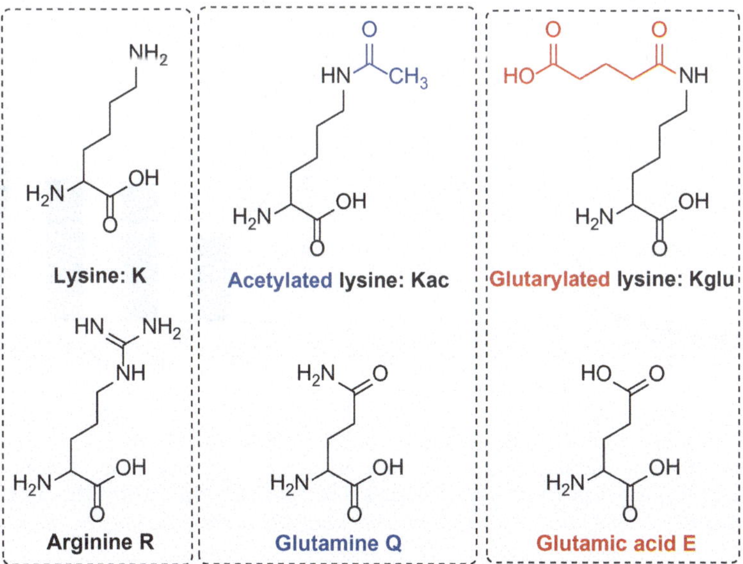

Fig. 4.3 Structures of Lysine (K), acetylated lysine (Kac), glutarylated lysine (Kglu) and their corresponding mimics, arginine (R), glutamine (Q), glutamic acid (E)

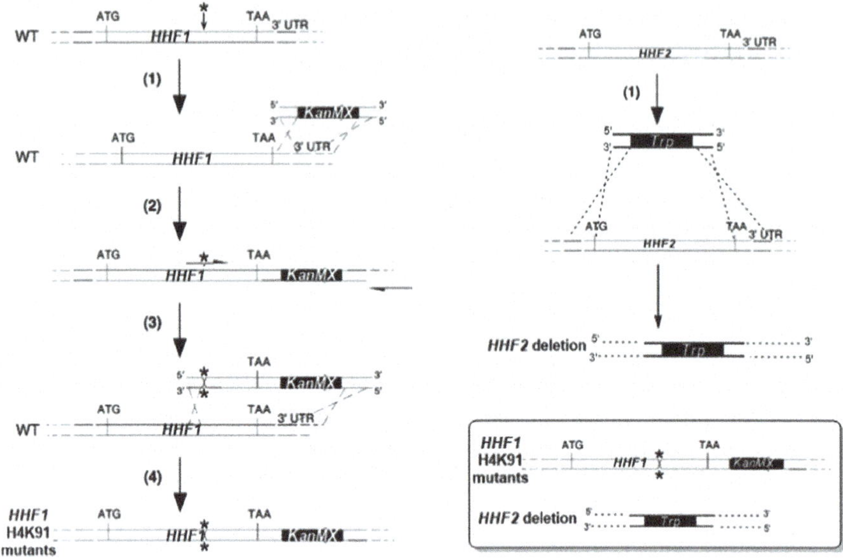

Fig. 4.4 Experimental strategy for construction of budding yeast strains containing H4K91 alleles

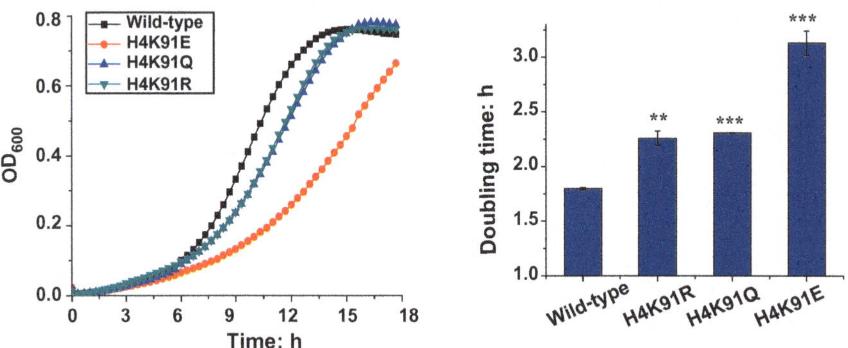

Fig. 4.5 Growth curve and quantitative analysis on cell cycle duration of wild-type cells and those containing H4K91E/Q/R alleles. Error bars indicated ± s.e. from three independent biological replicates. The p values are based on the Student's t test. $*p < 0.05$, $**p < 0.01$, $***p < 0.001$ (Bao et al. 2019)

4.3 Effects of H4K91 on Cell Cycle Progress

To explore the biological functions of H4 Lys 91 glutarylation, we firstly examined cell growth pattern of wild-type and H4K91 alleles. As shown in Fig. 4.5, the H4K91R and H4K91Q mutations showed similar and moderate effects in impeding cell proliferation. The H4K91E mutation, mimicking constitutive glutarylation, conferred the strongest effects on cell cycle delay. The doubling time of H4K91E

Table 4.1 Strains used in this study

Strain	Genotype
YPH499 (Wild type)	*MAT a met15-Δ0 ura3-52 lys2-801 ade2-101 his3-Δ200 trp1-Δ63 leu2-Δ1*
WYYY411	*MAT a met15-Δ0 ura3-52 lys2-801 ade2-101 his3-Δ200 trp1-Δ63 leu2-Δ1, hhf2Δ::TRP1 hhf1::KanMX6*
WYYY390	*MAT a met15-Δ0 ura3-52 lys2-801 ade2-101 his3-Δ200 trp1-Δ63 leu2-Δ1, hhf2Δ::TRP1 hhf1-K91E::KanMX6*
WYYY394	*MAT a met15-Δ0 ura3-52 lys2-801 ade2-101 his3-Δ200 trp1-Δ63 leu2-Δ1, hhf2Δ::TRP1 hhf1-K91Q::KanMX6*
WYYY398	*MAT a met15-Δ0 ura3-52 lys2-801 ade2-101 his3-Δ200 trp1-Δ63 leu2-Δ1, hhf2Δ::TRP1 hhf1-K91R::KanMX6*
WYYY383	*MAT a met15-Δ0 ura3-52 lys2-801 ade2-101 his3-Δ200 trp1-Δ63 leu2-Δ1, hhf2Δ::TRP1*
WYYY362	*MAT a met15-Δ0 ura3-52 lys2-801 ade2-101 his3-Δ200 trp1-Δ63 leu2-Δ1, hhf1::KanMX6*
WYYY376	*MAT a met15-Δ0 ura3-52 lys2-801 ade2-101 his3-Δ200 trp1-Δ63 leu2-Δ1, hhf1-K91E::KanMX6*
WYYY378	*MAT a met15-Δ0 ura3-52 lys2-801 ade2-101 his3-Δ200 trp1-Δ63 leu2-Δ1, hhf1-K91Q::KanMX6*
WYYY380	*MAT a met15-Δ0 ura3-52 lys2-801 ade2-101 his3-Δ200 trp1-Δ63 leu2-Δ1, hhf1-K91R::KanMX6*

cells was extended from 1.8 h (wild-type cells) to 3.3 h. These results indicate that modification at H4 Lys 91 is crucial for cell proliferation. It should be noted that the phenotypes observed in the presence of the H4K91 alleles were not due to alterations in histone H4 protein expression levels (Fig. 4.6).

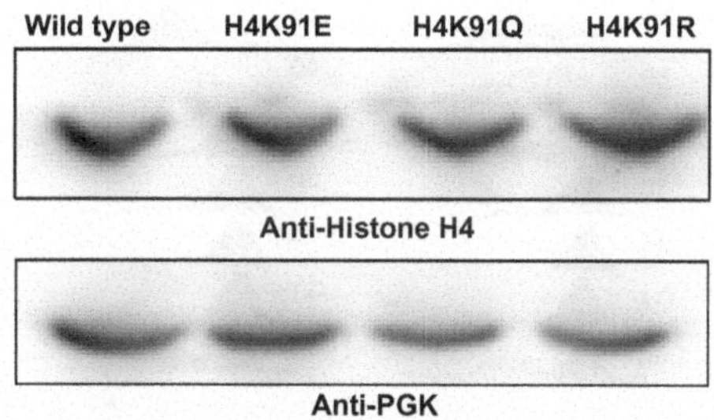

Fig. 4.6 Histone H4 expression. Immunoblotting analysis on extracted yeast whole-cell lysate using anti-histone H4 and anti-PGK antibodies, respectively (Bao et al. 2019)

4.4 Effects of H4K91 on DNA Replication

There are two distinct steps of DNA replication involving dramatic changes in chromatin structure, transient disruption of nucleosomes ahead of replication forks and nucleosome reassembly on nascent DNA [13, 14]. Both processes are modulated by various histone modifications [9]. We reasoned that the substitution of positively charged lysine by negatively charged glutamate could interrupt the form of salt bridge between H3/H4 tetramer and H2A/H2B dimers. The H4K91E mutation may therefore promote nucleosome deassembly but prevent nucleosome reassembly. The delayed cell cycle progression may partially result from failure in completion of DNA replication. To test this hypothesis, we performed cell cycle analysis of both wild-type and mutant cells. As we expected, Fig. 4.7 showed that H4K91E significantly delayed the completion of DNA replication from 60 min to 90 min, with stronger effect even compared to lysine acetylation mimic H4K91Q. The delayed S-phase progression mediated by H4K91E mutant indicated that the de novo assembly of chromatin, commonly accompanied by a large amount of histone synthesis during replication, could be compromised.

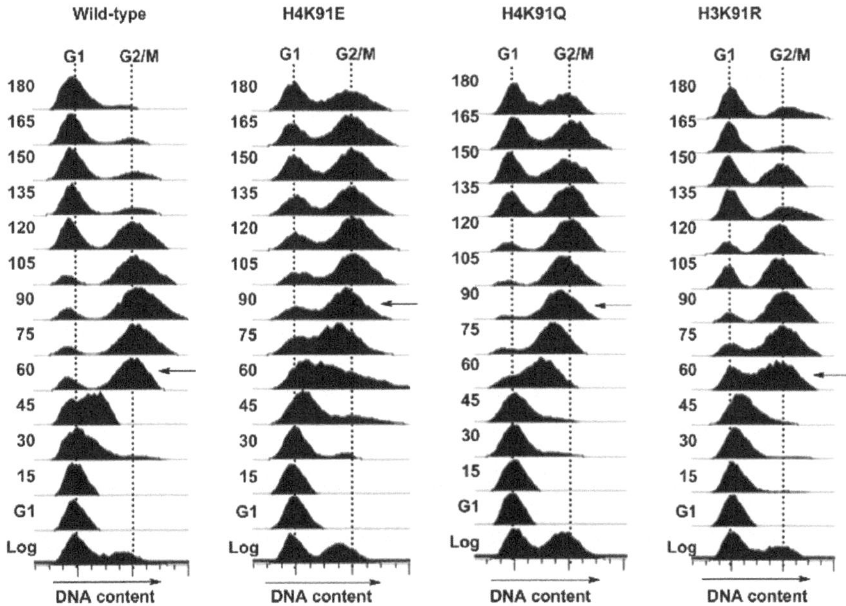

Fig. 4.7 Cell cycle analysis of wild-type cells and strains containing H4K91E/Q/R alleles via flow cytometry. Cells were synchronized into G1 phase via incubation with α-factor for 3 h and then collected every 15 min after releasing for analysis. DNA content at indicated times was determined via propidium iodide (PI) staining

4.5 Influence of H4K91 on DNA Damage Repair

Given that sensitivity to DNA damaging agents is one of the phenotypes often observed in strains containing defective chromatin assembly factors [15], the viability of both the wild-type cell and cells containing H4K91 alleles were assayed upon challenges from a series of DNA damaging agents, including methyl methanesulfonate (MMS, Figs. 4.9 and 4.11), hydroxyurea (HU, Figs. 4.10 and 4.11) and UV irradiation (Fig. 4.12). MMS, a DNA alkylating agent, can modify both guanine (to 7-methylguanine) and adenine (to 3-methlyladenine) to cause base mispairing and replication blocks [16]. Treatment with HU, an inhibitor of ribonucleotide reductase (RNR), can cause depletion of deoxynucleotide (dNTP) in cells, which results in replication fork arrest [17]. UV radiation may lead to mutagenic DNA lesions by inducing interstrand cross-linking [18]. In consistent to analysis result of growth curve that H4K91 mutants delayed the growth rate but not led to cell death, there was no difference in cell viability among all the strains tested after about 24 h incubation (Fig. 4.8). However, upon treatment with external DNA damaging agent, MMS, only H4K91E mutation caused detectable increase in DNA damage sensitivity (Figs. 4.9 and 4.11). For exposure to HU and UV radiation, both H4K91E and H4K91Q mutants, mimicking lysine glutarylation and acetylation, respectively, were markedly more sensitive than H4K91R and wild-type (Figs. 4.10, 4.11 and 4.12).

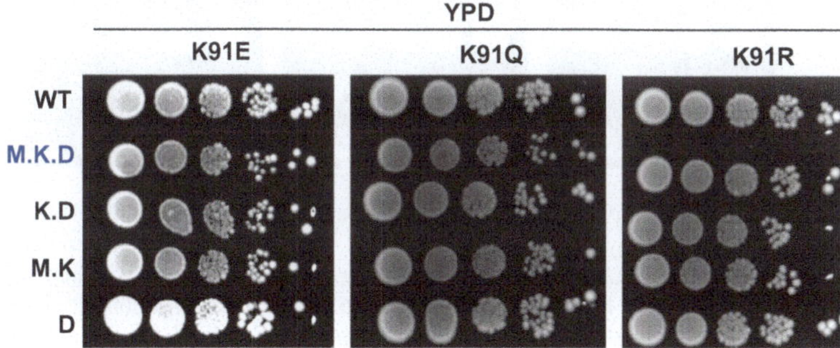

Fig. 4.8 Growth of wild-type cells and those containing H4K91E/Q/R alleles on normal yeast extract-peptone-dextrose (YPD) plates. 10-fold serial dilutions of cells, containing the indicated allele of histone H4, were plated on yeast YPD plates. WT: Wild-type, M.K.D: *HHF1* K91mutation with *KanMX6* marker, *HHF2* deletion with *Trp* marker, K.D: *HHF1* with *KanMX6* marker, *HHF2* deletion with *Trp* marker, M.K: *HHF1* K91 mutation with *KanMX6* marker, D: *HHF2* deletion with *Trp* marker

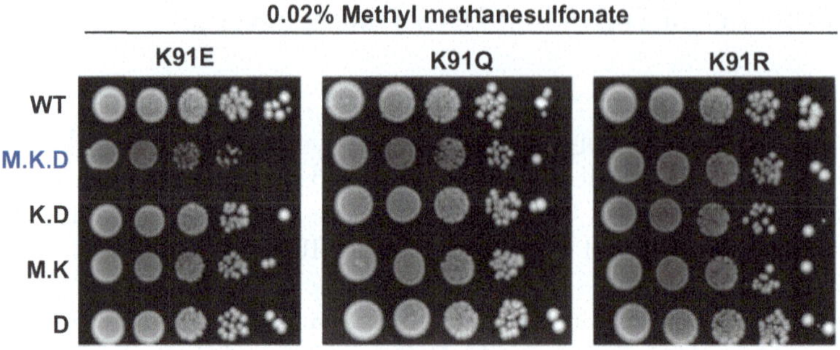

Fig. 4.9 DNA damage sensitivity assessment of wild-type cells and those containing H4K91E/Q/R alleles upon treatment with methyl methanesulfonate (MMS). 10-fold serial dilutions of cells, containing the indicated allele of histone H4, were plated on plate ± the indicated DNA damaging agent. WT: Wild-type, M.K.D: *HHF1* K91mutation with *KanMX6* marker, *HHF2* deletion with *Trp* marker, K.D: *HHF1* with *KanMX6* marker, *HHF2*deletion with *Trp* marker, M.K: *HHF1* K91mutation with *KanMX6* marker, D: *HHF2* deletion with *Trp* marker

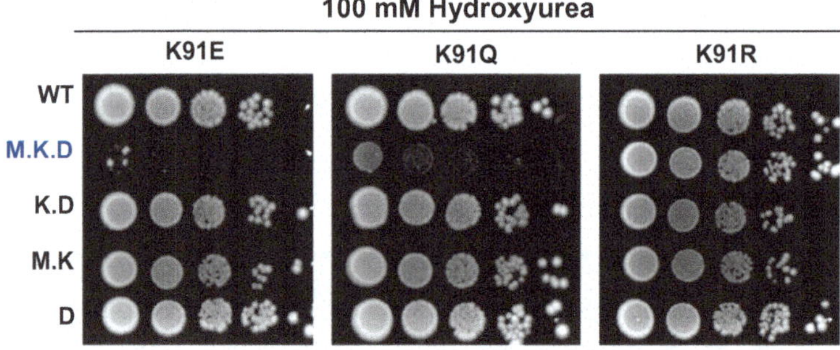

Fig. 4.10 DNA damage sensitivity assessment of wild-type cells and those containing H4K91E/Q/R alleles upon treatment with hydroxyurea (HU). 10-fold serial dilutions of cells, containing the indicated allele of histone H4, were plated on YPD plate ± the indicated DNA damaging agent. WT: Wild-type, M.K.D: *HHF1* K91mutation with *KanMX6* marker, *HHF2* deletion with Trp marker, K.D: *HHF1* with *KanMX6* marker, *HHF2*deletion with *Trp* marker, M.K: *HHF1* K91mutation with *KanMX6* marker, D: *HHF2* deletion with *Trp* marker

4.6 Genetic Evidence on How H4K91 Mutation Affects DNA Damage Repair

To further elucidate the role of H4K91glu in chromatin assembly process, we generated mutants of factors as to distinct aspects of DNA repairing process. As shown in Fig. 4.13, H4K91E mutation exacerbated the HU sensitivity of stains lacking *YKU70* (nonhomologous end-joining) or *RAD52* (homologous recombination), indicating that histone H4 Lys 91 was not directly involved in the DNA repair process. The HU sensitivity of strains that with depletion of *MEC1*, which is a critical kinase for the

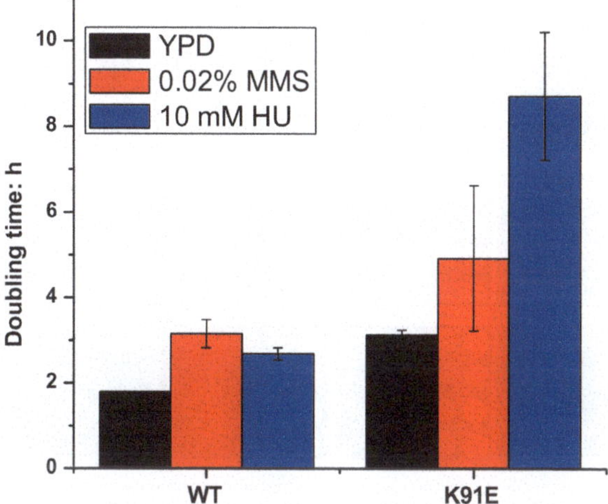

Fig. 4.11 Quantitative analysis on cell cycle duration of wild-type cells and those containing H4K91E allele in the presence of 0.02% MMS or 100 mM of HU. Error bars indicated ±s.e. from three independent biological replicates

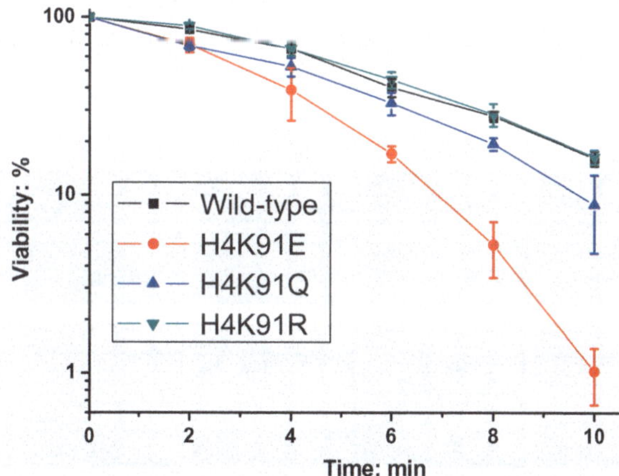

Fig. 4.12 UV sensitivity analysis. Cells were plated on YPD plates and then subjected to UV radiation for indicated time. Error bars indicated ±s.e. from three independent biological replicates

DNA damage checkpoint activation, was not increased by histone H4K91E allele. This result suggested the potential association between H4K91 glutarylation and DNA damage checkpoint activation. In addition, the H4K91E allele did not aggravate HU sensitivity when combined with factors involved in chromatin assembly. More specifically, mutants with gene deletion of *ASF1*, a histone chaperone, *CAC1*, a subunit of CAF-1 chromatin complex, or *HIR1*, a subunit of nucleosome complex, were sensitive to HU treatment, and the sensitivity was not deteriorated by addition

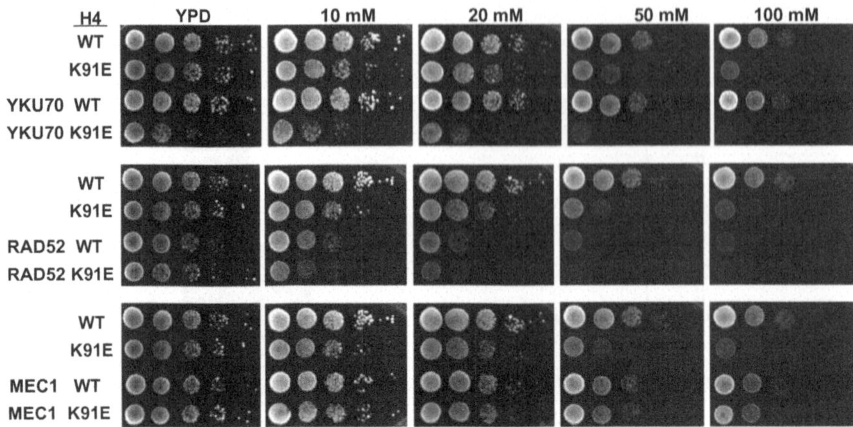

Fig. 4.13 The indicated histone H4 alleles were introduced into strains deleted for factors involved in non-homologous end joining (*YKU70*), homologous recombinational repair (*RAD52*), S-phase checkpoint (*MEC1*). Tenfold dilutions of each strain were analyzed in YPD plate or plates containing the indicated concentration of hydroxyurea. (Bao et al. 2019)

of H4K91E mutation (Fig. 4.14). These genetic data further reveal the involvement of H4K91 in chromatin assembly modulation that occurs in the repair process of damaged DNA.

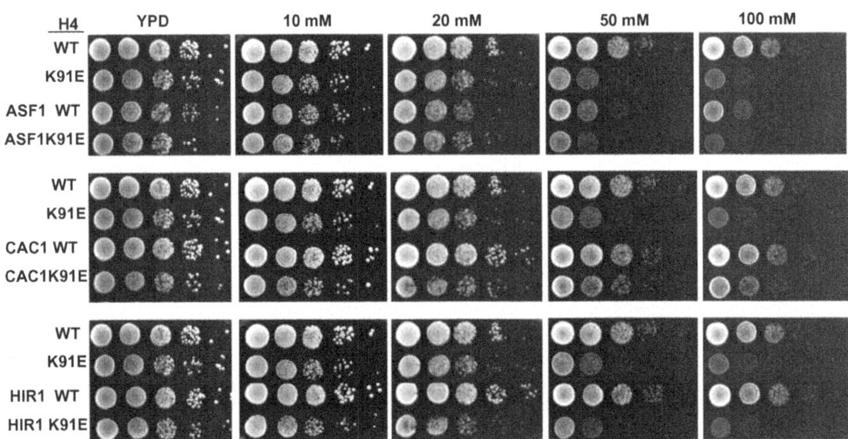

Fig. 4.14 The indicated histone H4 alleles were introduced into strains with depletion of factors *ASF1* (histone chaperone), *CAC1* (a subunit of chromatin assembly complex CAF-1) and *HIR1* (a subunit of nucleosome assembly complex). Tenfold dilutions of each strain were analyzed in YPD plate or plates containing the indicated concentration of hydroxyurea. (Bao et al. 2019)

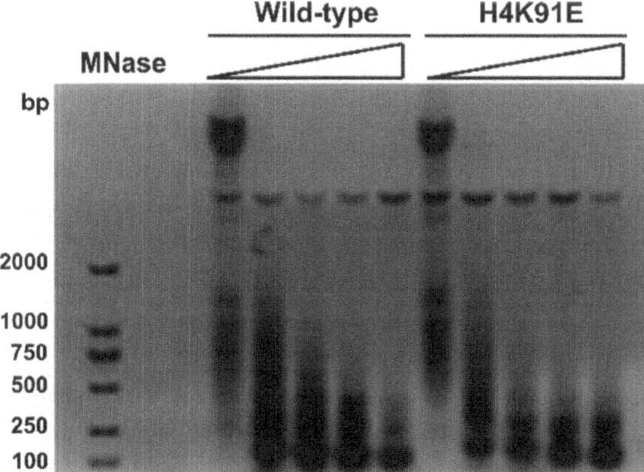

Fig. 4.15 Micrococcal nuclease (MNase) digestion assay. Equal quantities of nuclei from wild-type and H4K91E cells were digested with MNase for the indicated times. After deproteination, DNA was analyzed by 1% agarose gel electrophoresis. bp: base pair

4.7 Relaxed Chromatin Compaction State Resulted from H4K91E Mutation

The defects in chromatin assembly would likely lead to global defects in chromatin structure. Therefore, to obtain direct evidence that H4K91glu modulates the chromatin assembly, chromatin from both wild-type and H4K91E cells were isolated and subjected to digestion by micrococcal nuclease (MNase). The accessibility of nuclease would be augmented if compaction of chromatin were loosed. Consistent with loss of intranucleosomal histone-histone interactions, chromatins extracted from H4K91E cells were digested more rapidly relative to wild-type ones, indicating that histone H4K91E indeed could weaken chromatin stability via interrupting histone–histone interactions (Fig. 4.15).

4.8 Discussion

For understanding the role of Kglu in regulation of chromatin dynamics, we turn to glutarylation at histone H4 lysine 91. Unlike modifications at histone 'tails', H4K91 lies in the interface between H2A/H2B dimer and H3/H4 tetramer. The positively charged Lys 91 can form salt bridge with histone H2B Glu 68 and thereby is essential to maintain the stability of nucleosome. The interruption of the salt bridge is therefore expected to promote nucleosome deassembly and prevent nucleosome assembly. To assess the role of glutarylation at H4K91, we generated yeast strains with mutation from K91 to Glu, Gln and Arg. The genetic data showed that the substitution of Lys 91 to Glu resulted in delayed cell cycle progression and enhanced sensitivity to exogenous DNA damaging agents. One mechanism to explain those phenotypes caused by H4K91E mutation was that the presence of negatively charged carboxylate could disrupt the formation of salt-bridge between histones and lead to destabilized nucleosome structure.

In addition to glutarylation, H4K91 undergoes other modifications, such as acetylation [19] and ubiquitination [20]. Histone H4 protein copurified with the histone acetyltransferase Hif1p complex was revealed to be modified by acetylation at lysine 91. The association of acetylated histone H4 and the nuclear Hat1p-Hat2p-Hif1p complex was consistent with the presence of this acetylation on newly synthesized molecules. A model was therefore proposed for the role of H4K91 acetylation in the nucleosome assembly that the acetylation at K91 occurs on newly synthesized histone H4 prior to assembly onto DNA and after deposition of the acetylated H3/H4 tetramer onto DNA, this acetyl moiety should be removed for the incorporation of H2A/H2B dimers and the completion of nucleosome assembly. BBAP, an E3 ligase for protein ubiquitination, was revealed to selectively modify H4K91 with monoubiquitination and protect cells from external DNA damages. The BBAP-mediated ubiquitination at lysine 91 was associated with methylation at H4 Lys 20, which was docking site for 53BP1, and therefore affect the kinetics of 53BP1 foci formation at DNA damaged sites. It was hypothesized that the addition of a relatively large molecule, ubiquitin, to the globular core domain could impact the chromatin structure to expose H4K20 for posttranslational modifications.

Recently, studies of H2AX revealed that its ubiquitination requires its prior acetylation and the sequential acetylation and ubiquitination of H2A.X promote enhanced histone dynamics to stimulate DNA damage response [21]. Although the mechanistic association between H4K91 acetylation and ubiquitination has been elusive, an increase of H4K91 acetylation was observed in the absence of H4K91 ubiquitination [20]. However, H3K79 methylation, which is involved in recruitment of 53BP1 [22], was revealed function independent on H4K91 acetylation [19]. In our study, H4K91E, a mimic of lysine glutarylation, showed stronger sensitivity to all the tested DNA damage agents than H4K91Q, mimicking lysine acetylation. In addition, H4K91A allele was reported to increase DNA damage sensitivity of strains with depletion of *MEC1* [19], but the sensitivity was not accentuated by H4K91E mutation. Collectively, it is certain that chemical modifications, including acetylation, glutarylation and ubiquitination, at histone H4 Lys 91 are involved in the regulation of nucleosome assembly, but the subsequent complicated signaling pathways in chromatin-templated nuclear process have not been well defined.

4.9 Experimental Methods

Yeast Strains Construction

Yeast genetic manipulations were done according to standard methods. Histone alleles were generated via two steps: I) deletion of unnecessary *HHF2* gene with *TRP* marker and II) site-directed mutation from lysine (K) to glutamic acid (E), glutamine (Q) and arginine (R) in *HHF1* with incorporation of *KanMX6* at C-terminal. *MEC1*, *RAD53*, *YKU70*, *RAD52*, *SWC5*, *ASF1*, *CAC1* and *HIR1* were replaced by *HisMX6* marker.

Growth Curve Analysis

The yeast strains were grown exponentially in YPD at 30 °C with continuous rotation. Cells were collected by centrifugation, resuspended in sterile water to 0.2×10^7 cells/mL. Each well of 96-well non-coated polystyrene microplates containing 95 μL of fresh YPD media with or without corresponding concentration of MMS or HU was inoculated with 5 μL of yeast cultures (5 × 104 cells). The samples were prepared in triplicate and cell growth was monitored with microplate scanning spectrophotometer at 600 nm every 20 min during a 24 h period. The whole set of experiments were repeated at least three times.

Flow Cytometry

Yeast cells are synchronized into G1 phase via incubation with α-factor for 3 h and then collected every 15 min after releasing for analysis. DNA content at indicated times will be determined via flow cytometric analysis. Cells at different time points are collected by centrifugation and fixed in 70% ethanol for overnight. After centrifugation, the cell pellet will be washed and then treated with RNase A (1 mg/ml) and proteinase K (1 mg/ml) for 1 h at 37 °C and 55 °C respectively. Samples will be stained with 5 μg/mL propidium iodide at 4 °C overnight in the dark. A flow cytometer will be used to analyze the samples. For each histogram, 10,000 cells are assayed.

DNA Damage Assays

To test the effects of modifications at H4K91 on DNA damage sensitivity, both the WT and H4K91E/Q/R cells were treated with MMS, HU and UV irradiation. Briefly, serial dilutions of the cells were made and spotted onto yeast extract–peptone–dextrose (YPD) media plates containing either MMS or HU. For UV sensitivity, cells were plated in triplicate on YPD plates and irradiated with a series of UV intensities. The viability of each strain was determined as the ratio of colonies formed on the UV-irradiated plates to those without UV exposure.

Chromatin Extraction from Yeast Cell Culture

500 mL of exponentially growing cells were collected and washed with sorbitol wash buffer (1.4 M Sorbitol, 40 mM HEPES, pH 7.5, 0.2 mM $MgCl_2$, 1 mM PMSF, 10 mM beta-ME). The pelleted cells were resuspended with 4 mL sorbitol digestion buffer (1.4 M Sorbitol, 40 mM HEPES, pH 7.5, 0.2 mM $MgCl_2$, 1 mM PMSF, 2 mM beta-mercaptoethanol) incubated in 37 °C water bath for 30 min with gentle shaking. Zymolase (20 mg/mL) was then added to a final concentration 0.5 mg/mL with another 30 min incubation to get spheroplast. After centrifugation, cell pellet was resuspended in ice-cold Ficoll buffer (18% (w/v) Ficoll 400, 20 mM PIPES pH 6.5, and 0.5 mM $MgCl_2$.) and transferred to a prechilled glass homogenizer with

homogenization on ice using pestle A for 30 even strokes. The homogenate was then gently layered onto ice-cold glycerol–Ficoll buffer (7% (w/v) Ficoll 400, 20% (v/v) glycerol, 20 mM PIPES–Na pH 6.5, and 0.5 mM $MgCl_2$) in a round-bottom 50-mL tube. The nuclei were pelleted through the glycerol–Ficoll cushion by centrifugation at 21,500 × g for 30 min at 4 °C and resuspended again with Ficoll buffer. The debris and intact cells were removed by centrifugation at 3300 × g for 15 min at 4 °C and the supernatant was centrifuged again at 21,500 × g for 30 min at 4 °C to collect final chromatin.

MNase Digestion

Extracted nuclei were preincubated at 37 °C and then digested with a series of concentration of MNase at 37 °C for 10 min. Reactions were stopped by adding 0.5 M EDTA. Following RNase A and proteinase K treatment to remove RNA and proteins contained in nuclei, DNA was isolated via phenol-chloroform-isoamyl alcohol extraction. Ethanol precipitated DNA was washed with 70% ethanol, dissolved in ddH_2O, electrophoresed in 1.6% agarose gel and visualized by ethidium bromide staining.

References

1. Tan S, Davey CA (2011) Nucleosome structural studies current opinion in structural biology 21:128–136. https://doi.org/10.1016/j.sbi.2010.11.006
2. Richmond TJ, Davey CA (2003) The structure of DNA in the nucleosome core. Nature 423:145–150. https://doi.org/10.1038/nature01595
3. Widom J (1998) Structure, dynamics, and function of chromatin in vitro. Ann Rev Biophys Biomol Struct 27:285–327
4. Luger K (2006) Dynamic nucleosomes. Chrom Res Int J Mol Supramol Evol Asp Chrom Biol 14:5–16. https://doi.org/10.1007/s10577-005-1026-1
5. Ehrenhofer-Murray AE (2004) Chromatin dynamics at DNA replication, transcription and repair. Eur J Biochem FEBS 271:2335–2349. https://doi.org/10.1111/j.1432-1033.2004.04162.x
6. Groth A, Rocha W, Verreault A, Almouzni G (2007) Chromatin challenges during DNA replication and repair. Cell 128:721–733. https://doi.org/10.1016/j.cell.2007.01.030
7. Zentner GE, Henikoff S (2013) Regulation of nucleosome dynamics by histone modifications. Nat Struct Mol Biol 20:259–266. https://doi.org/10.1038/nsmb.2470
8. Bannister AJ, Kouzarides T (2011) Regulation of chromatin by histone modifications. Cell Res 21:381–395. https://doi.org/10.1038/cr.2011.22
9. Kouzarides T (2007) Chromatin modifications and their function. Cell 128:693–705. https://doi.org/10.1016/j.cell.2007.02.005
10. Bowman GD, Poirier MG (2015) Post-translational modifications of histones that influence nucleosome dynamics. Chem Rev 115:2274–2295. https://doi.org/10.1021/cr500350x
11. Cross NC (2012) Histone modification defects in developmental disorders and cancer. Oncotarget 3:3–4. https://doi.org/10.18632/oncotarget.436
12. Xie Z et al (2012) Lysine succinylation and lysine malonylation in histones. Mol Cell Proteomics: MCP 11:100–107. https://doi.org/10.1074/mcp.M111.015875
13. Demeret C, Vassetzky Y, Mechali M (2001) Chromatin remodelling and DNA replication: from nucleosomes to loop domains. Oncogene 20:3086–3093. https://doi.org/10.1038/sj.onc.1204333

14. Falbo KB, Shen X (2006) Chromatin remodeling in DNA replication. J Cell Biochem 97:684–689. https://doi.org/10.1002/jcb.20752
15. Adams CR, Kamakaka RT (1999) Chromatin assembly: biochemical identities and genetic redundancy. Curr Opin Genet Devel 9:185–190. https://doi.org/10.1016/S0959-437X(99)80028-8
16. Beranek DT (1990) Distribution of methyl and ethyl adducts following alkylation with monofunctional alkylating agents. Mutat Res 231:11–3017
17. Koc A, Wheeler LJ, Mathews CK, Merrill GF (2004) Hydroxyurea arrests DNA replication by a mechanism that preserves basal dNTP pools. J Biol Chem 279:223–230. https://doi.org/10.1074/jbc.M303952200
18. Rastogi RP, Richa, Kumar A, Tyagi MB, Sinha RP (2010) Molecular mechanisms of ultraviolet radiation-induced DNA damage and repair. J Nucleic Acids 2010:592980. https://doi.org/10.4061/2010/592980
19. Ye J et al (2005) Histone H4 lysine 91 acetylation a core domain modification associated with chromatin assembly. Mol Cell 18:123–130. https://doi.org/10.1016/j.molcel.2005.02.031
20. Yan Q, Dutt S, Xu R, Graves K, Juszczynski P, Manis JP, Shipp MA (2009) BBAP monoubiquitylates histone H4 at lysine 91 and selectively modulates the DNA damage response. Mol Cell 36:110–120. https://doi.org/10.1016/j.molcel.2009.08.019
21. Ikura T et al (2007) DNA damage-dependent acetylation and ubiquitination of H2AX enhances chromatin dynamics. Mol Cell Biol 27:7028–7040. https://doi.org/10.1128/mcb.00579-07
22. Huyen Y et al (2004) Methylated lysine 79 of histone H3 targets 53BP1 to DNA double-strand breaks. Nature 432:406–411. https://doi.org/10.1038/nature03114

Chapter 5
Identification of Sirt3 as an 'Eraser' for Histone Lysine Crotonylation Marks Using a Chemical Proteomics Approach

5.1 Introduction

Increasing evidences have indicated histone posttranslational modifications (PTMs) play crucial roles in regulating a wide range of chromatin-templated nuclear processes, such as gene transcription, DNA replication and DNA damage repair [1]. PTMs on histones have been proposed to serve as histone code that provides epigenetic information from a mother cell to daughter cells [2]. Histone code is under control of opposing enzymes that 'write' or 'erase' modifications on histones [1, 3]. Meanwhile, this epigenetic code could be translated by 'reader' proteins, which can recognize specific histone modifications, into biological readout, without alteration of genetic information [4, 5].

Lysine crotonylation (Kcr) is newly identified as a histone PTM that is specifically marks active gene promoters and potential enhancers in mammalian cell genomes [6]. This modification is specifically enriched at testis-specific X-linked genes that are essential for male germ cell differentiation in postmeiotic male germ cells [7]. However, lack of knowledge about enzymes that catalyze the incorporation or elimination of Kcr in cells greatly hinders further mechanism and function studies on histone Kcr.

Based on a systematic activity screening of eleven human zinc-dependent lysine deacetylases (HDAC1 to HDAC11) [8] and a radioactive thin layer chromatography (TLC)-based assay [9], HDAC3 in complex with nuclear receptor corepressor 1 (HDAC3-NCoR1) and Sirt1-2 were identified to show decrotonylation activity, respectively. However, those discoveries only relied on activity determination against a series of C-terminal lysine acylated peptides in a fluorometric assay or a histone H3K9Cr peptide in radioactive [^{32}P]-NAD thin layer chromatography assay. In addition, their catalytic mechanisms and the molecular basis of substrate recognition are still unknown. It is therefore still essential to identify and characterize endogenous histone decrotonylases. To fill this knowledge gap, we need a method to profile 'eraser' enzymes that recognize histone Kcr.

X. Bao, *Study on the Cellular Regulation and Function of Lysine Malonylation, Glutarylation and Crotonylation*, Springer Theses,
https://doi.org/10.1007/978-981-15-2509-4_5

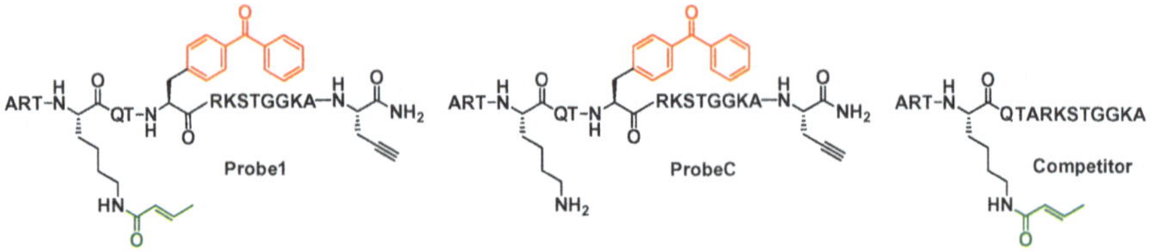

Fig. 5.1 Chemical structures of probe **1**, probe **C** and competitor H3K4Cr. [22]

5.2 Chemical Proteomics Approach for Identification Binding Proteins of Histone H3K4Cr Mark

A cross-linking-assisted and SILAC-based protein identification (CLASPI) approach has been recently reported to identify histone PTM 'readers' [10, 11]. We reasoned that this approach could also be used for capture of PTM 'erasers', since the preliminary step for PTM 'erasers' to execute their enzymatic activity is to recognize and bind their specific PTM substrates. In addition, the short lifetime of photoactive groups endows the CLASPI approach with the capacity to capture of transient and/or weak interactions. Here, we optimized CLASPI approach to comprehensively profile 'eraser' enzymes that recognize histone Kcr marks.

5.2.1 Peptide Probe Design

We first focused on a crotonylation mark discovered on histone H3K4 [6] and designed a peptide probe (probe **1,** Fig. 5.1) based on the unstructured *N*-terminal region of histone H3, with Lys 4 crotonylated. A photo-cross-linker (benzophenone) was appended to alanine-7 to convert the non-covalent protein-protein interactions, mediated by this Kcr mark, into irreversible covalent linkages via its photoactivity. In addition, a bio-orthogonal function group (alkyne) is added at the peptide *C*-terminus to enable selective isolation of captured binding partners. As a negative control, probe **C** was designed to be identical to probe **1** except for Lys 4 crotonylation.

5.2.2 Strategy of Cross-Linking Assisted and SILAC Based Protein Identification (CLASPI) Approach

To identify proteins that bind H3K4Cr with high selectivity and tight affinity, we performed two types of CLASPI experiments with cell lysates derived from HeLa S3 cells grown in 'heavy' (^{13}C, ^{15}N-substitued arginine and lysine) or 'light' medium (Fig. 5.2). For selectivity filtration, the 'heavy' and 'light' cell lysates were incubated with probe **1** and probe **C**, respectively, for photoactive reaction and then pooled

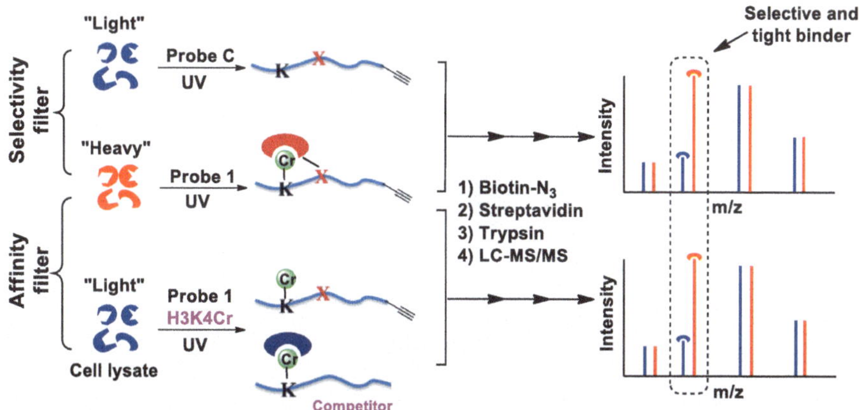

Fig. 5.2 Schematic diagram illustrating the CLASPI strategy to profile proteins that bind H3K4Cr with high selectivity and affinity in whole-cell proteomes. LC-MS, liquid chromatography–mass spectrometry. [22]

together. The later conjugation with a biotin tag via click chemistry contributed to the isolation of those linked proteins. After in-gel trypsin digestion, the tryptic peptide mixtures were separated by HPLC and analyzed with a LTQ-Orbitrap mass spectrometer. In this way, the H3K4Cr-selective binding proteins are likely to show a high H/L ratio. To further distinguish the high-affinity interactions, both lysates were photo-cross-linked with probe **1**, but a competitor at concentration of 30 μM was also contained in the 'light' sample (Fig. 5.2). We speculated that H3K4Cr binding proteins that have high-affinity (K_d < 30 μM) will be stand out with a high H/L ratio from complex proteome. Together, a protein showing high H/L ratio in both experiments will be considered as a selective and tight binder of H3K4Cr.

5.2.3 *Sirt1, Sirt2, and Sirt3 Recognize Histone H3K4Cr Mark*

A logarithmic (Log_2) SILAC ratio (H/L) of the identified proteins was then plotted with data from the 'selectivity-filter' experiment along the *x*-axis and 'affinity-filter' experiment along the *y*-axis. As shown in Fig. 5.3, three nicotinamide adenine dinucleotide (NAD)-dependent deacetylases [12–14], Sirt1-3, preferentially bond to this histone Kcr mark with more than 10-fold enrichment by crotonylated probe **1**, in contrast to the nonspecific binding proteins that showed SILAC (H/L) ratios around 1. Among these three selective H3K4Cr binders, Sirt3 appeared as an outlier outside of the background in the 'affinity-filter' experiment with the highest H/L. The representative mass spectrometry spectra for peptides from Sirt1, Sirt2 and Sirt3 in 'selectivity-filter' and 'affinity-filter' experiments were shown in Figs. 5.4 and 5.5, respectively. This result indicates that Sirt3 is likely a binding partner with selectivity and relatively tight affinity toward crotonylation mark at H3 Lys 4.

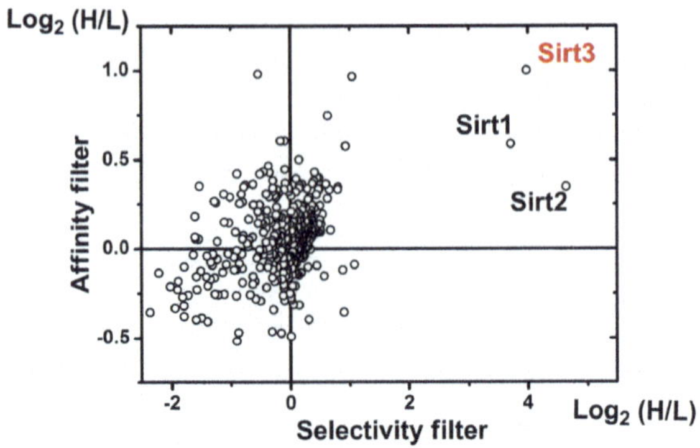

Fig. 5.3 A two-dimensional plot showing the Log_2 values of the stable isotope labeling of amino acids in cell culture (SILAC) ratios (heavy/light (H/L)) of each identified protein for the 'selectivity filter' (*x* axis) and 'affinity filter' (*y* axis) experiments. [22]

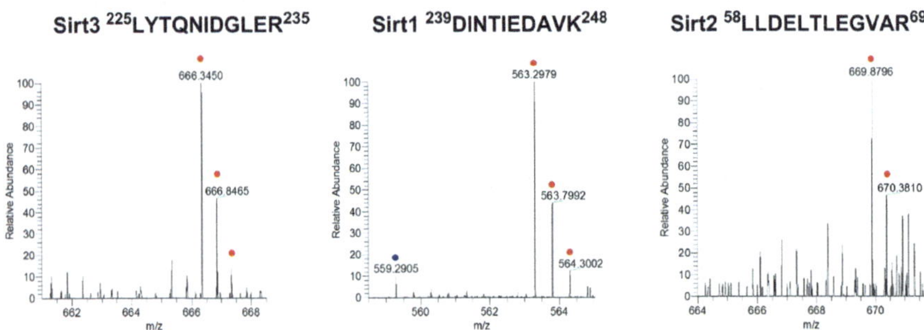

Fig. 5.4 Representative mass spectrometry spectra for peptides from Sirt1, Sirt2 and Sirt3, identified in 'selectivity-filter' CLASPI experiment. The 'light' and 'heavy' peptide isotopes are indicated by blue and red dots, respectively. [22]

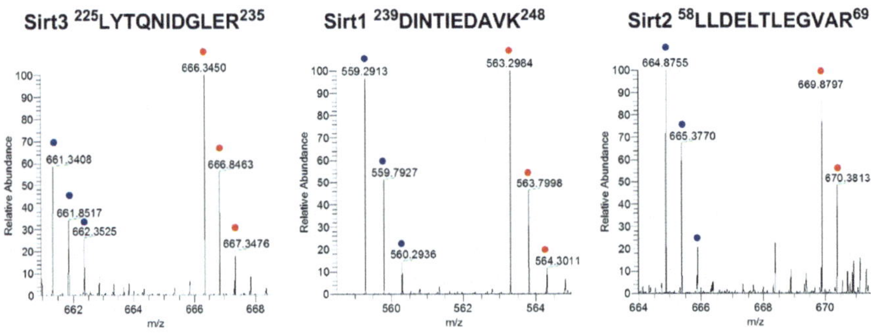

Fig. 5.5 Representative mass spectrometry spectra for peptides from Sirt1, Sirt2 and Sirt3, identified in 'affinity-filter' CLASPI experiment. The 'light' and 'heavy' peptide isotopes are indicated by blue and red dots, respectively. [22]

5.3 Expression and Purification of Human Sirt1, Sirt3, Sirt3 H248Y and Sirt5

To corroborate the result of CLASPI experiment, we next determined to examine the direct and selective binding to crotonylated histone peptide using recombinant human sirtuins. To this end, we firstly performed plasmid construction and/or protein purification of sirtuins, including Sirt1, Sirt3, Sirt3 H248Y and Sirt5. All the proteins used in this study were expressed in *Rosetta* cells with integrated His-tag. After two-step purification using Nickel-affinity column and gel-filtration column, we got the very pure target protein. Fractions were pooled together for assessment of enzymatic reactivity and binding affinity (Figs. 5.6, 5.7, 5.8 and 5.9).

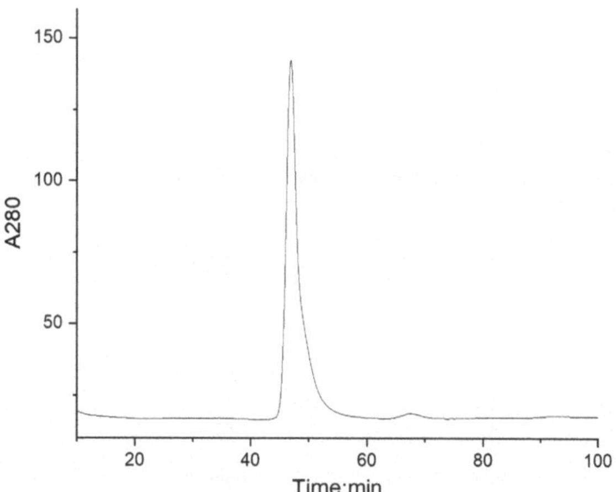

Fig. 5.6 Purification of recombinant humsn Sirt1. The purifty of Sirt1 was analyzed via gel-filtration chromatography

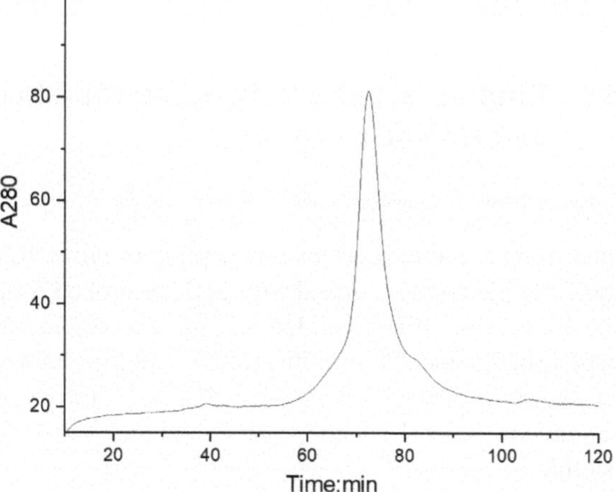

Fig. 5.7 Purification of recombinant humsn Sirt3. The purifty of Sirt3 was analyzed via gel-filtration chromatography

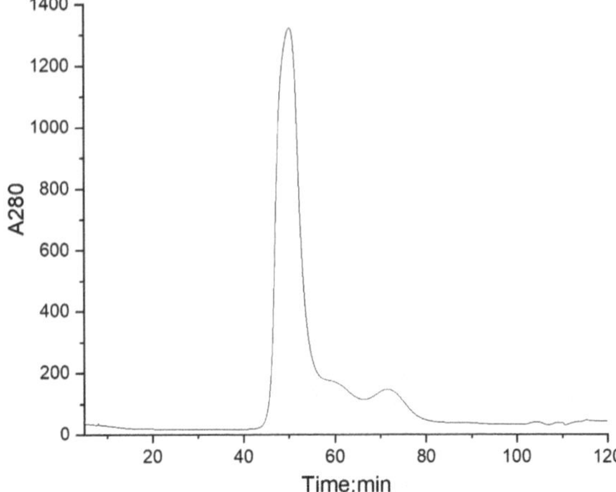

Fig. 5.8 Purification of recombinant humsn Sirt3 H248Y mutant. The purifty of Sirt3 H248Y was analyzed via gel-filtration chromatography

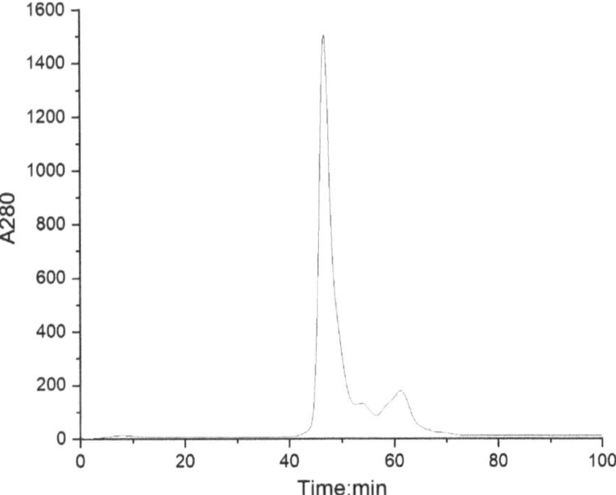

Fig. 5.9 Purification of recombinant humsn Sirt5. The purifty of Sirt5 was analyzed via gel-filtration chromatography

5.4 Binding Affinity Between Recombinant Sirt3 and H3K4Cr Peptide

We next examined whether recombinant human Sirt3 can directly and selectively bind to this crotonylated histone peptide in vitro. To do this, 20 ng/μL of purified Sirt3 was photo-cross-linked with 2 μM of probe **C** or probe **1** to validate the binding specificity. For affinity validation, Sirt3 was pre-incubated with 30 μM of H3K4Cr peptide before incubation with probe **1**. Single protein without any treatment was used as a negative control. As shown in Fig. 5.10, the recombinant Sirt3 was selectively captured by probe **1** but not probe **C**, and the cross-linking was inhibited by H3K4Cr peptide.

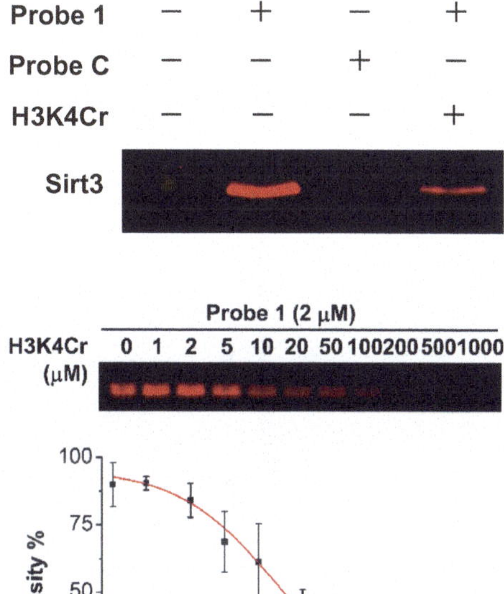

Fig. 5.10 Recombinant Sirt3 was selectively labeled in vitro by crotonylated probe **1** (2 μM) and the labeling by probe **1** was inhibited by H3K4Cr peptide (30 μM). [22]

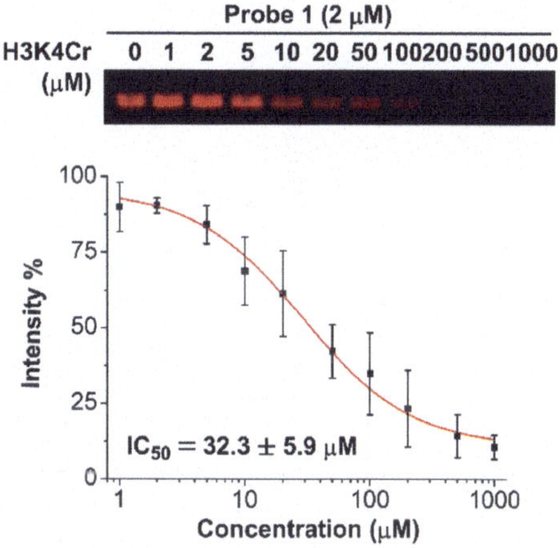

Fig. 5.11 Determination of IC_{50} for inhibition of probe **1**-induced labeling of Sirt3 by H3K4Cr peptide ($n = 3$, mean ± s.e.). [22]

To further confirm the high binding affinity between Sirt3 and H3K4Cr, we performed competition assay, in which, in addition to 2 μM of probe **1**, a series of concentration of competitor H3K4Cr peptide was added to interfere photoactive interaction between Sirt3 and probe **1.** In consistent with CLASPI result, the specific binding was inhibited with an IC_{50} value = 32.3 μM (Fig. 5.11).

5.5 Decrotonylation Activity of Sirtuins

5.5.1 NAD-Dependent Decrotonylation Activity

Inspired by the fact that Sirt3 was identified as a NAD-dependent deacetylase, we set out to investigate whether Sirt3 catalyzes NAD-dependent decrotonylation reaction. LC-MS was used to monitor the hydrolysis of the H3K4Cr peptide. As expected, the removal of crotonyl group only occurred in the presence of both Sirt3 and NAD (Fig. 5.12a, b). The conclusion that Sirt3 catalyzes decrotonylation in NAD-dependent manner was also supported by the complete loss of decrotonylase activity of Sirt3 H248Y (Fig. 5.12c). Histidine 248 was revealed as a conserved and active residue in sirtuins for their deacetylation activity. Denu et al. demonstrated

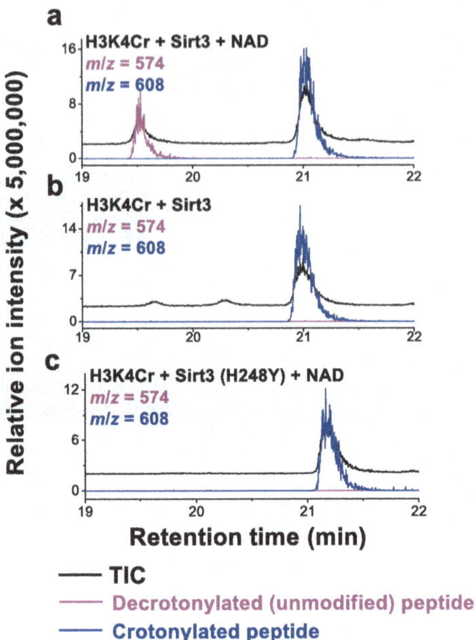

Fig. 5.12 Sirt3 catalyzes the hydrolysis of crotonyllysine in vitro. The hydrolysis of the crotonylated peptides by Sirt3 was analyzed by LC-MS. The hydrolysis of H3K4Cr was observed with Sirt3 in the presence (**a**) but not absence of NAD (**b**) or with Sirt3 H248Y mutant (**c**). Black traces show total ion intensity for all ion species with m/z from 300 to 2000 (i.e., total ion counts, TIC); pink traces show ion intensity (5 × magnified) for the masses of decrotonylated (unmodified) peptides; and blue traces show ion intensity (5 × magnified) for the masses of crotonylated peptides. [22]

that mutation of this histidine led to not only compromised deacetylation but also absence of nicotinamide, a product from NAD. In addition, *O*-crotonyl-adenosine 5′-diphosphoribose (*O*-Cr-ADPR) as a product of this hydrolysis reaction was detected by LC-MS (Fig. 5.13), further demonstrating that Sirt3 catalyzes the hydrolysis of crotonyllysine in the same way as it hydrolyzes acetyllysine [15] (Fig. 5.14).

5.5.2 *Sequence-Dependent Decrotonylation Activity*

We further assessed the decrotonylation activity of Sirt3 toward a collection of crotonylated histone peptides, H3K9Cr, H2BK5Cr, H3K27Cr and H3K8Cr. As shown in Fig. 5.15, Sirt3 manifested varied decrotonylation activities toward these peptides, indicating the sequence-dependent decrotonylation activity of Sirt3.

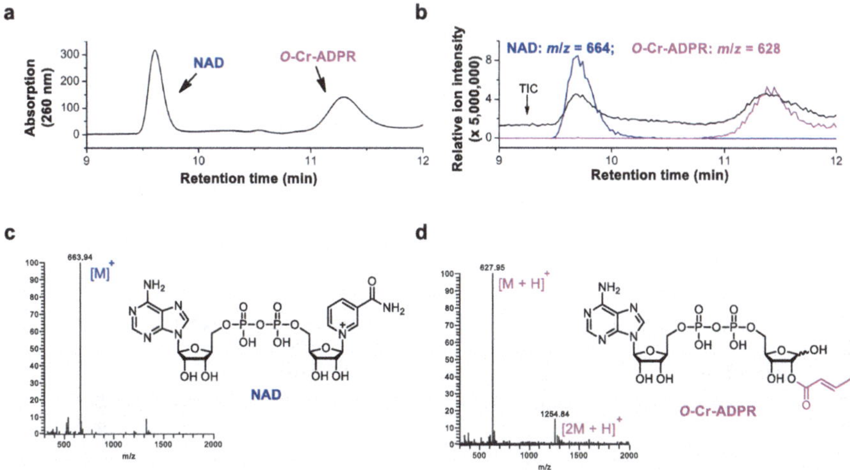

Fig. 5.13 Detection of *O*-crotonyl-adenosine 5′-diphosphoribose (*O*-Cr-ADPR) by LC-MS. The reaction mixture of Sirt3 catalyzed NAD-dependent decrotonylation of H3K4Cr peptide was analyzed by LC-MS. Partial chromatogram of the reaction mixture with UV (260 nm) (**a**) and mass detector (**b**). In (**b**), the black trace shows total ion intensity for all ion species with m/z from 300 to 2000 (i.e., total ion counts, TIC); blue trace shows ion intensity (5 × magnified) for the mass of NAD (m/z = 664); and pink trace shows ion intensity (5 × magnified) for the mass of *O*-Cr-ADPR (m/z = 628). (**c**) ESI-MS spectra of NAD. (**d**) ESI-MS spectra of *O*-Cr-ADPR. [22]

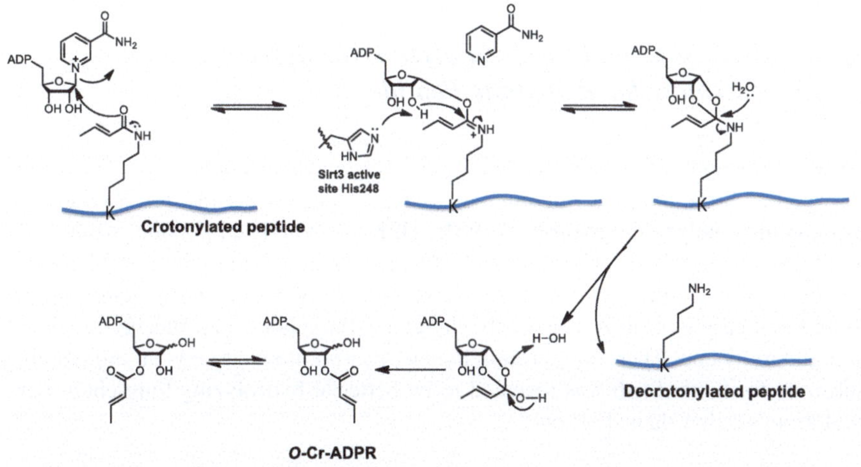

Fig. 5.14 Proposed mechanism of Sirt3-catalyzed NAD-dependent decrotonylation. [22]

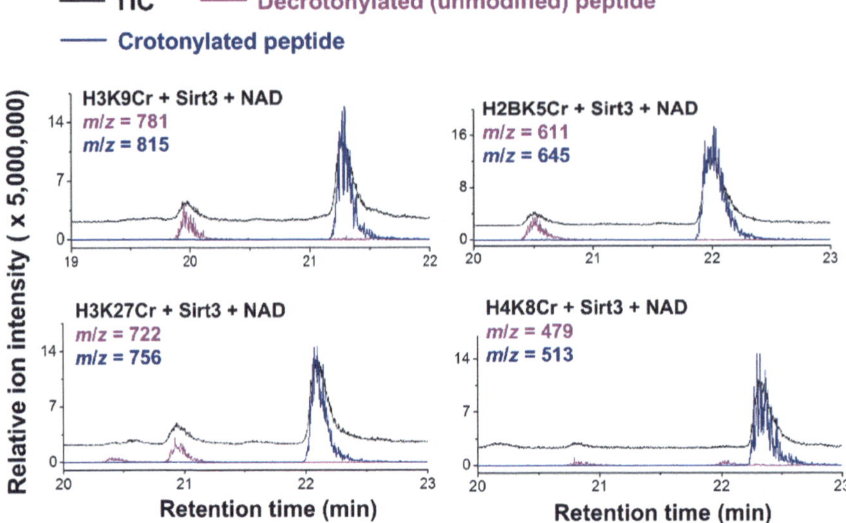

Fig. 5.15 Sirt3 catalyzes the hydrolysis of crotonyllysine in vitro. Sirt3 showed varied decrotonylation activities towards H2BK5Cr, H3K9Cr, H3K27Cr, and H4K8Cr peptides. Black traces show total ion intensity for all ion species with m/z from 300 to 2000 (i.e., total ion counts, TIC); pink traces show ion intensity (5 × magnified) for the masses of decrotonylated (unmodified) peptides; and blue traces show ion intensity (5 × magnified) for the masses of crotonylated peptides. [22]

5.5.3 Sirt1, Sirt2 and Sirt3 Catalyze Hydrolysis of Crotonylated Histone Peptide

We also investigated the decrotonylation activity of other sirtuin proteins. The results showed that Sirt1 and Sirt2, although as weaker binders toward H3K4Cr relative to Sirt3, could catalyze the hydrolysis of the H3K4 crotonylated peptide, while Sirt5 and Sirt6 showed weak or undetectable decrotonylation activity (Fig. 5.16). Sirt5, one of the four sirtuins with weak deacetylase activity, was recently demonstrated to preferentially hydrolyze negatively charged carboxylates, like succinyl, malonyl and glutaryllysine. Therefore, crotonyllysine, bearing no charge, is an unfavorable substrate for Sirt5. Sirt6 was revealed to be better at hydrolyzing long-chain fatty acyl groups rather than short ones.

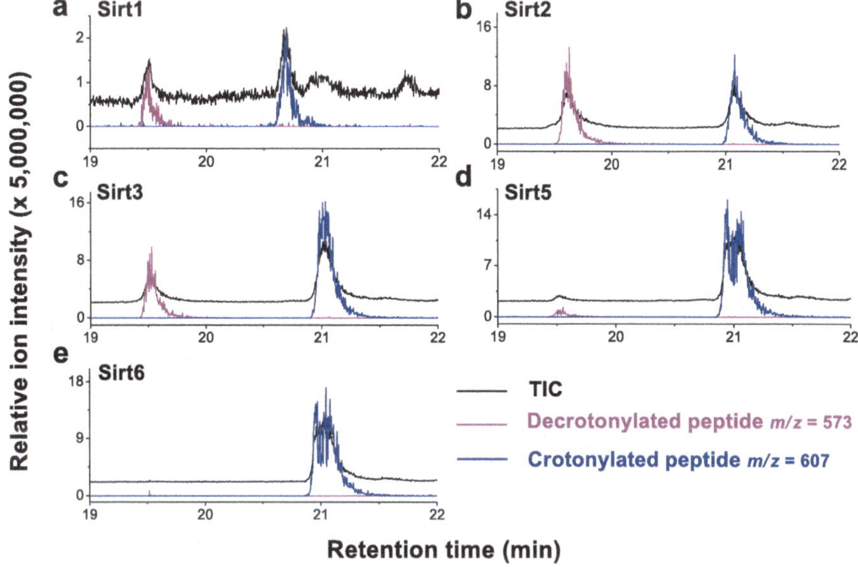

Fig. 5.16 Decrotonylation activity of sirtuins in vitro. The hydrolysis of the H3K4Cr peptide by sirtuins was analyzed by LC-MS. **a–c** Sirt1, Sirt2, and Sirt3 showed significant decrotonylation activities towards H3K4Cr. **d**, **e** Sirt5 and Sirt6 showed little decrotonylation activities towards H3K4Cr. Black traces show total ion intensity for all ion species with m/z from 300 to 2000 (i.e., total ion counts, TIC); pink traces show ion intensity (5 × magnified) for masses of decrotonylated (unmodified) peptides; and blue traces show ion intensity (5 × magnified) for masses of crotonylated peptides. [22]

5.6 Direct Determination of Binding Affinity Between Sirtuins and Crotoylated Peptides Using Isothermal Titration Calorimetry

5.6.1 Binding Affinity Between Sirt1-3 and H3K4Cr Peptide

In this section, isothermal titration calorimetry (ITC), a direct measurement of binding affinity, was carried out to determine the binding affinity between sirtuins and crotonylated histone peptides. Careful analysis revealed the K_d values for Sirt2 and Sirt3 toward H3K4Cr were 95.2 μM and 25.1 μM, respectively. While the interaction between Sirt1 and H3K4Cr was so weak that K_d value was not determined via this method (Fig. 5.17). The results, in accordance with result of mass spectrometry analysis, showed that Sirt1 and Sirt2 had lower affinities toward the H3K4Cr peptide, further indicating that they are selective but relatively weak binders for histone H3K4 crotonylation mark.

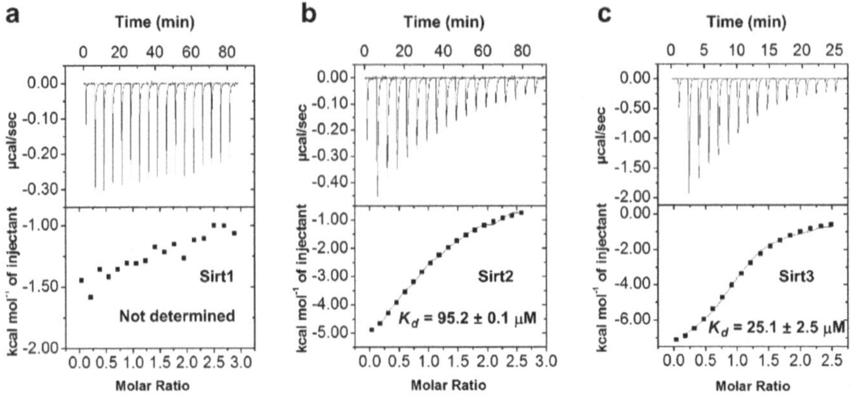

Fig. 5.17 Isothermal titration calorimetry measurement for the binding affinity of Sirt1 (**a**), Sirt2 (**b**) and Sirt3 (**c**) towards the H3K4Cr peptide. [22]

5.6.2 *Binding Affinity Between Sirt3 and Crotonylated Histone Peptides*

We further assessed the binding affinity of Sirt3 toward those crotonylated histone peptides that were used in enzymatic assay. As shown in Fig. 5.18, Sirt3 showed varied binding affinity towards H3K9Cr and H3K27Cr with K_d values equal to 24.75 μM and 156.99 μM, respectively, but weak interactions with H2BK5Cr and H4K8Cr. The varied binding affinities of Sirt3 to these peptides can partially explain the observation that Sirt3 showed substrate selectivity as a decrotonylase.

5.6.3 *Crystal Structure Analysis of Sirt3-H3K4Cr Complex*

Crystallographic analysis of Sirt3 in complex with H3K4Cr peptide (PDB 4V1C, done by Dr. Yi Wang) reveals that H3K4Cr is bound in a similar way to the published complex structure of Sirt3 with a lysine acetylated AceCS2 peptide (PDB 3GLR) [16] (Fig. 5.19). Residue His 248, a catalytic residue for Sirt3 as a deacetylase, forms a hydrogen bond with the crotonyl amide oxygen. In addition to His 248, hydrophobic residues Phe 180, Ile 230, Ile 291 and Phe294 of Sirt3 forms a binding pocket for crotonyllysine (Fig. 5.20). This discovery further validated the NAD-dependent mechanism of Sirt3-catalyzed decrotonylation. The extensive hydrophobic and hydrogen-bonding interactions between Sirt3 and the peptide side chains in Sirt3-H3K4Cr complex structure (PDB 4V1C) can partially explain the sequence-dependent decrotonylation activity and binding affinity of Sirt3 toward a collection of lysine crotonylated histone peptides. Strikingly, a robust π-π stacking

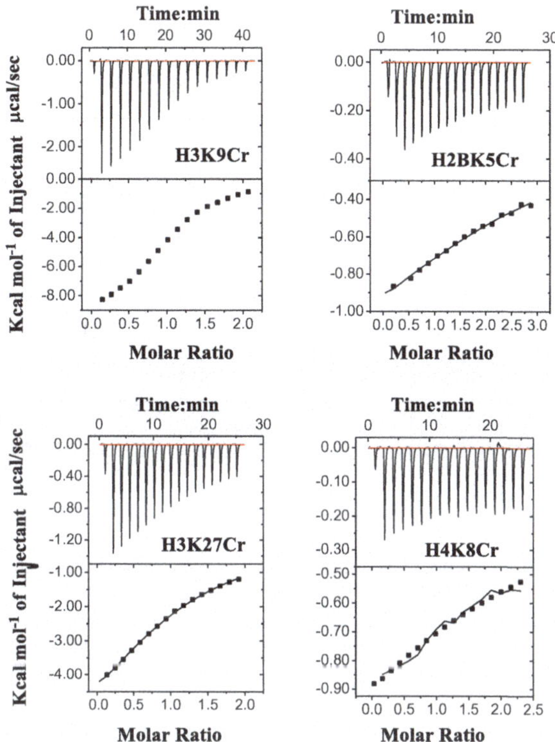

Fig. 5.18 Isothermal titration calorimetry measurement for the binding affinities of Sirt3 towards H3K9Cr (24.75 µM), H2BK5Cr (ND), H3K27Cr (156.99 µM) and H4K8Cr (ND). ND: not determined. [22]

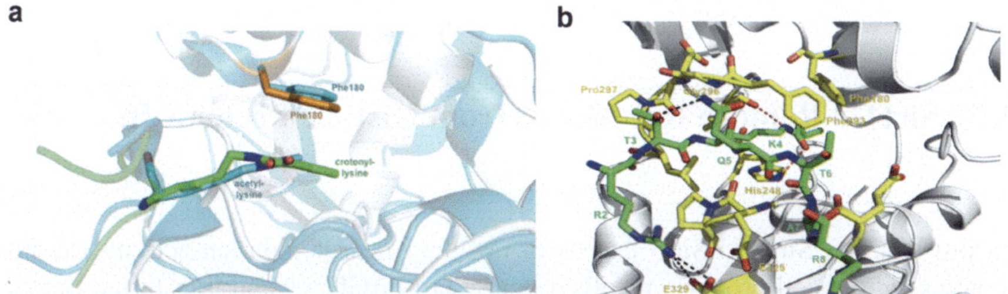

Fig. 5.19 Detailed structural analysis for Sirt3 in complex with H3K4Cr peptide. **a** Structural alignment between the Sirt3–AceCS2 complex colored in cyan (PDB 3GLR) and the Sirt3–H3K4Cr complex colored in orange. **b** Interactions between Sirt3 and H3K4Cr peptide. The residues of Sirt3 are colored in yellow and H3K4Cr peptide is colored in green. [22]

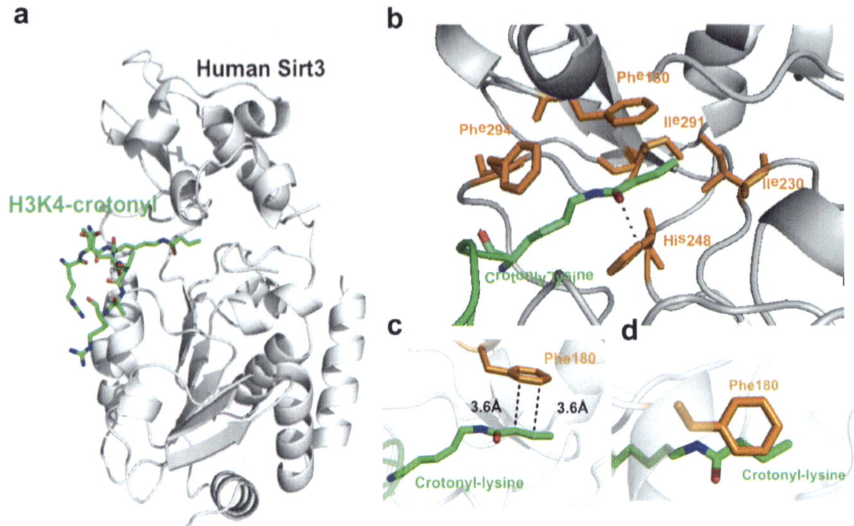

Fig. 5.20 Structural basis for how Sirt3 recognizes lysine crotonylation. **a** Overall structure of the complex of Sirt3 (gray) with H3K4Cr peptide (green). **b** The binding pocket formed by hydrophobic residues (orange) that accommodate the crotonyllysine. A side view **c** and top view **d** of a $\pi - \pi$ stacking interaction between residue Phe180 of Sirt3 and the crotonyl group. [22]

interaction occurs between phenyl ring of residue Phe180 and carbon-carbon double bond (C = C) in crotonyl group. A primary sequence alignment of all sirtuins revealed that the phenylalanine residue (Phe180 of Sirt3) is only conserved in Sirt1-3, but not in other sirtuins (Fig. 5.21). This is why Sirt4–Sirt7 were not identified in our CLASPI experiments.

5.7 Sirt1-3 Remove Kcr Marks from Histone Proteins In Vitro

To test whether Sirt1 to Sirt3 can carry out their decrotonylation activity toward proteins, whole-cell proteins, resolved in an SDS-PAGE gel and transferred onto a PVDF membrane were incubated with the enzymes, respectively, in the presence of NAD. The crotonylation level change detected a pan anti-Kcr antibody revealed substantial reductions on lysine crotonylation level on two bands with a molecular weight similar to histones (Fig. 5.22).

Given that Sirt1-3 can catalyze the hydrolysis of crotonylated histone peptides in vitro, we speculated that histones could be one of these proteins with Kcr level reduction. To test this hypothesis, core histones were extracted from HeLa cells for on-membrane decrotonylation experiment. Briefly, following separation by SDS-PAGE gel and transfer onto PVDF membrane, histones were incubated with Sirt1,

```
SIR1  197 DPRTILKDLLPET----IPPPELDDMTLWQIVINILSEPPKRKK--------RKDINTIEDAVKLLQ--ECKKIIVLTGAGVSVSCGIPDFRSR-DGIYARLAVDFPDLPDPQA  295
SIR2   54 ----------------------------------------QKER--------LLDELTLEGVARYMQSERCRRVICLVGAGISTSAGIPDFRSPSTGLYDNLEK--YHLPYPEA  117
SIR3  107 ----------------------------------------GASSVVGSGGSSDKGKLSLQDVAELIRARACQRVVVMVGAGISTPSGIPDFRSPGSGLYSNLQQ--YDLPYPEA  178
SIR4   31 ---------LFVP----ASP--------------------------------PLDPEKVKELQRFIT--LSKRLLVMTGAGISTESGIPDYRSEKVGLYARTDR--RPIQHGDF   95
SIR5   20 KPPA--------------STR-------NQICL---------KM--------ARPSSSMADFRKFFA--KAKHIVIISGAGVSAESGVPTFRGA-GGYWRKWQA--QDLATPLA   90
SIR6   20 ---------LPE----IFDP-----------------------P--------EELERKVWELARLVW--QSSSVVFHTGAGISTASGIPDFRGP-HGVWTMEER---------   76
SIR7   71 ---------LKRRQEEVCDD-----------------------P--------EELRGKVRELASAVR--NAKYLVVYTGAGISTAASIPDYRGP-NGVWTLLQK---------  131

           180                                                      230             248
SIR1  296 MFDIEYFRKDPRP--FFK--FAK-EIYPGQFQPSLCHKFIALSD----KEGKLLRNYTQNIDTLEQVAGIQ--RIIQCHGSFATASCLIC--K--YKVD------CE--AVRG-  385
SIR2  118 IFEISYFKKHPEP--FFA--LAK-ELYPGQFKPTICHYFMRLLK----DKGLLLRCYTQNIDTLERIAGLEQEDLVEAHGTFYTSHCVSASCR--HEYP------LS--WMKE-  211
SIR3  179 IFELPFFFHNPKP--FFT--LAK-ELYPGNYKPNVTHYFLRLLH----DKGLLLRLYTQNIDGLERVSGIPASKLVEAHGTFASATCTVC--Q--RPFP------GE--DIRA-  270
SIR4   96 V------RSAPIRQRYWARNFVG-WPQFSSHQPNPAHWALSTWE----KLGKLYWLVTQNVDALHTKAGSR--RLTELHGCMDRVLCLDC--G--EQTPR------G--VLQER  184
SIR5   91 ------FAHNPSRVWEFY--HYR-REVMGSKEPNAGHRAIAECETRLGKQGRRVVVITQNIDELHRKAGTK--NLLEIHGSLFKTRCTSC--G--VVAENYKSPICP--ALSG-  186
SIR6   77 ---------------GLA--PKF-DTTFESARPTQTHMALVQLE----RVGLLRFLVSQNVDGLHVRSGFPRDKLAELHGNMFVEECAKC----KTQYV------RD--TVVG-  155
SIR7  132 ---------------GRS--VSAADLS-E-AEPTLTHMSITRLH----EQKLVQHVVSQNCDGLHLRSGLPRTAISELHGNMYIEVCTSC--VPNREYV------RVFDVTER-  213

                                                               291 294
SIR1  386 -------------------D-------IFNQV-VPRCP--------RCPADEPLAIMKPEIVFFGEN--L--PEQFHRAMKYDKDEVDLLIVIGSSLKVRPVA-LIPS-S--I-  455
SIR2  212 -------------------K-------IFSEV-TPKCE--------DCQ-----SLVKPDIVFFGES--L--PARFFSCMQSDFLKVDLLLVMGTSLQVQPFA-SLIS-K--A-  276
SIR3  271 -------------------D-------VMADR-VPRCP--------VCT-----GVVKPDIVFFGEP--L--PQRFL-LHVVDFPMADLLLILGTSLEVEPFA-SLTE-A--V-  334
SIR4  185 FQVLNPTWSAEAHGLAPDGDVFLSEEQVRSFQ-VPTCV--------QCG-----GHLKPDVVFFGDT--V--NPDKVDFVHKRVKEADSLLVVGSSLQVYSGY-RFILTA--W-  276
SIR5  187 ----------KG---APEPGTQDASIPVEKLP-RCEEA--------GCG-----GLLRPHVVWFGEN--L--DPAILEEVDRELAHCDLCLVVGTSSVVYPAA-MFAP-Q--VA  265
SIR6  156 -------------------T-------MGLKATGRLCTVAKARGLRACR-----GELRDTILDWEDS--L--PDRDLALADEASRNADLSITLGTSLQIRPSG-NLPL-ATK--  230
SIR7  214 -----------------T-AL------HRHQ-TGRTCH--------KCG-----TQLRDTIVHFGERGTLGQPLN-WEAATEAASRADTILCLGSSLKVLKKYPRLWC-MTKP-  286

SIR1  456 PH-EVPQILINREPLP---------------------H-LHFDVELLGDCDVIINELCH--RLGGEYAKLC--CNPVKLSEITEKPPRTQKELAYLSELPPTPLHVSEDSSSPE  542
SIR2  277 PL-STPRLLINKEKAGQSDPFLGMIMGLGGGMDFDSKK-AYRDVAWLGECDQGCLALAE--LLGWKKELE---DLVRREHA-------------SIDAQSGAGVPNPSTSASPK  370
SIR3  335 RS-SVPRLLINRDLVG------------P---LAWHP-RSRDVAQLGDVVHGVESLVE--LLGWTEEMR---DLVQRET----------------------------------  391
SIR4  277 EK-KLPIAILNIGPTR---------------------SDDLACLKLNSRCGELLPLID-------------------------------------------------------  312
SIR5  266 AR-GVPVAEFNTETTP---------------------ATNRFRFHFQGPCGTTLPEALACHE---------------------------------------------------  305
SIR6  231 -RRGGRLVIVNLQPTK---------------------HDRHADLRIHGYVDEVMTRLMK--HLGLEIPAW---DGPRVLE----------RALPPLP-RPPTPKLEP-KEESPT  305
SIR7  287 PSRRPKLYIVNLQWTP---------------------KDDWAALKLHGKCDDVMRLLMA--ELGLEIPAYSRWQDP-----------------IFS--LATPL---------  347
```

Fig. 5.21 Sequence alignment of human Sirt1–Sirt7. The included sequences are Sirt1-197-542 (full length, 747 amino acids (a.a.)), Sirt2-54-370 (389 a.a.), Sirt3-107-391 (399 a.a.), Sirt4-31-312 (314 a.a.), Sirt5-20-305 (310 a.a.), Sirt6-20-305 (355 a.a.), and Sirt7-71-347 (400 a.a.). The residues of Sirt3 that interact with crotonyllysine in the crystal structure of the Sirt3-H3K4Cr complex are indicated (▼) above the residues and labeled with the Sirt3 sequence number. Highlighted in green is the Phe180 of Sirt3, which is involved in recognition of the crotonyllysine via a $\pi - \pi$ stacking interaction, and is conserved in Sirt1 and Sirt2, but not in the other sirtuins. The alignment was done by T-coffee. [22]

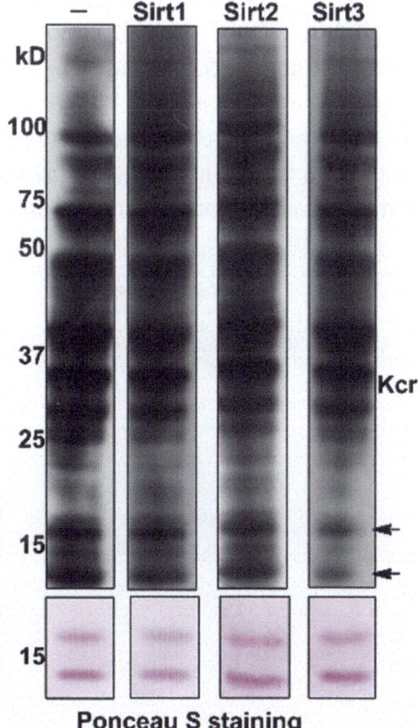

Fig. 5.22 Immunoblotting analyses showing Sirt1–3 catalyzed decrotonylation of whole-cell lysates on the membrane. The arrows indicate the protein bands with reduction of crotonylation levels. Ponceau S staining was used as the loading control. [22]

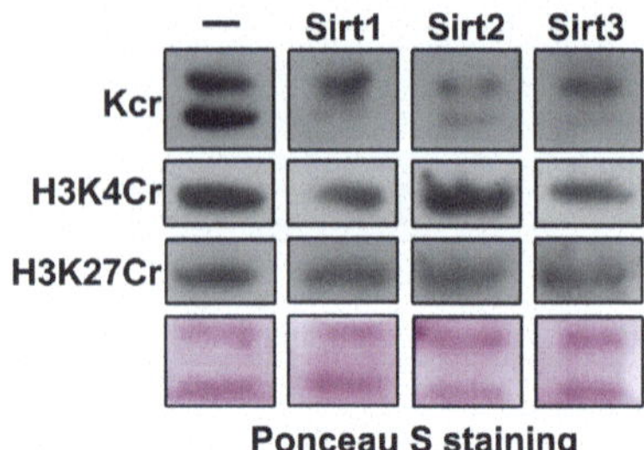

Fig. 5.23 Immunoblotting analyses showing Sirt1-3 catalyzed decrotonylation of extracted histones on membrane. Ponceau S staining was used as loading control. [22]

Sirt2 or Sirt3, separately. As a negative control, histones were treated only with NAD. As expected, Sirt1 to Sirt3 indeed reduced global Kcr level of all core histones extracted from HeLa cells and two known histone Kcr sites, H3K4Cr and H3K27Cr (Fig. 5.23).

5.8 Sirt3 Regulates Histone Kcr Levels in Living Cells

Finally, we examined the lysine crotonylation status on histones in HeLa cells, in which Sirt1, Sirt2 or Sirt3 was knockdown by siRNA, separately. As shown in Figs. 5.24 and 5.25, knockdowns of Sirt1 or Sirt2 by siRNA caused detectable augment on crotonylation level neither on global histones nor at the two tested Kcr sites, even though they can catalyze the removal of crotonyl from histone peptides and proteins in vitro. In contrast, downregulation of Sirt3 led to remarkable accumulation of lysine crotonylation not only on global histone but also at H3 Lys 4 residue.

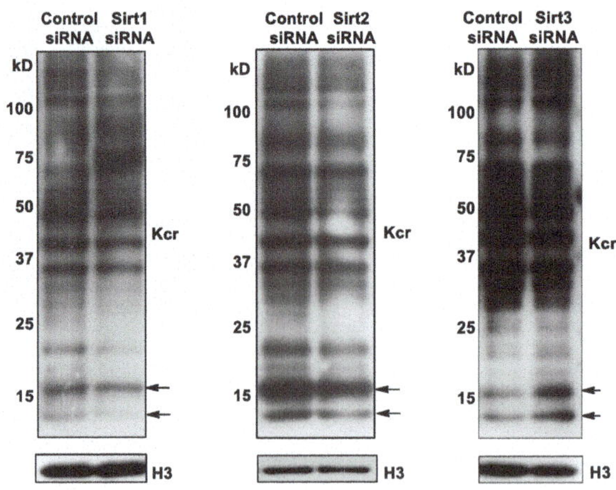

Fig. 5.24 Analysis of the decrotonylation activities of Sirt1-3 in cells. Immunoblotting analyses showing that Sirt3 knockdown but not Sirt1 or Sirt2 knockdown caused the accumulations in the global histone crotonylation level (as indicated by the arrows). H3 was used as loading control. [22]

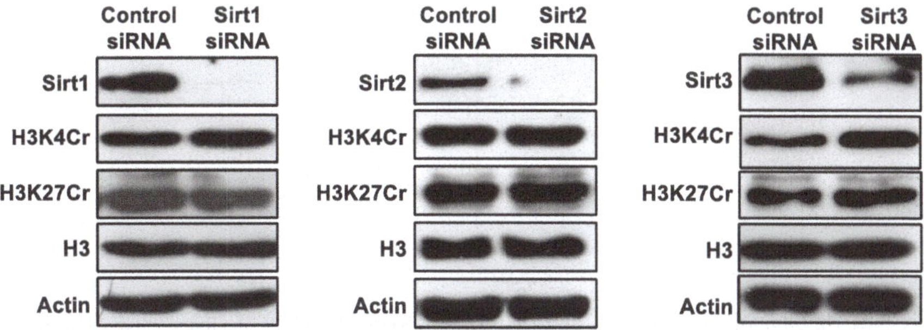

Fig. 5.25 Analysis of the decrotonylation activities of Sirt1-3 in cells. Immunoblotting analyses showing the influence of Sirt1-3 knockdown on H3K4Cr and H3K27Cr level in HeLa cells. H3 and actin were used as loading controls. [22]

Interestingly, in consistent with weaker activity of Sirt3 toward the H3K27Cr peptide in vitro, the altered Sirt3 expression level resulted in no appreciable effect on endogenous H3K27Cr level.

5.9 Subcellular Localization of Sirt3

The conclusion of Sirt3 as an endogenous histone decrotonylase is challenged by that Sirt3 was previously detected to be exclusively localized in mitochondria [17, 18] to involve in metabolic regulations through regulation on protein acetylation dynamics [19]. However, recent evidence has revealed the presence of full-length Sirt3 in the nucleus [20]. In support of this, using an antibody that specifically targets the *N*-terminal region of Sirt3, we indeed detected endogenous Sirt3 in the nucleus of HeLa cells by both immunofluorescence (Fig. 5.26) and immunoblotting analyses (Fig. 5.27). The nuclear localization of Sirt3 can also be supported the finding that Sirt3 can bind to chromatin and cause repression of the neighboring genes in U2OS

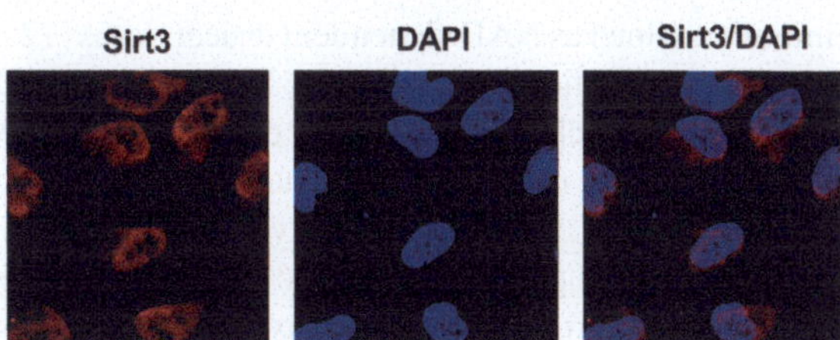

Fig. 5.26 Analysis of Sirt3 distribution in HeLa cells by fluorescence microscopy using anti-Sirt3 *N*-term antibody. Blue channel: DAPI. Red channel: Sirt3. [22]

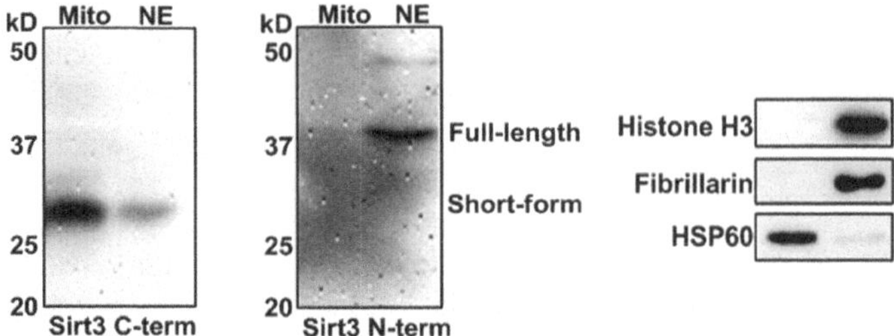

Fig. 5.27 Immunoblotting analyses showing the nuclear localization of endogenous Sirt3 using both anti-Sirt3 *C*-term and *N*-term antibodies. Immunoblotting analyses of fibrillarin/histone H3 and HSP60 were used to show the purity of nuclear (NE) and mitochondrial extraction (Mito), respectively. [22]

cells [21]. In addition, our labmate, Xiao-Meng Li observed significant increases in Kcr levels of five of the seven genes analyzed upon Sirt3 knockdown by siRNA, indicating that Sirt3 may directly regulate crotonylation dynamics at the genomic loci where it binds [22].

5.10 Discussion

We have established a robust chemical proteomics approach for profile of histone decrotonylases. The present studies have largely broadened the scope of CLASPI approach in identification of PTM-mediated protein-protein interactions, including not only stable protein–protein interactions between PTM and their 'readers' but also dynamic and transient interactions between PTMs and their 'erasers'. We found that Sirt1-3 can catalyze the hydrolysis of lysine crotonylated histone peptides and proteins in vitro. More importantly, we identified Sirt3 as an endogenous histone decrotonylase. This discovery provides new impetus to investigate the physiological significance of histone Kcr, but also helps to decipher the unknown cellular mechanisms controlled by Sirt3.

Sirtuins were initially defined as NAD-dependent deacetylases [12–14]. However, acetylation-independent deacylation activities have been detected in Sirt2, a robust deacetylase, Sirt5 and Sirt6, which displayed weak deacetylation activity. Sirt2 [23, 24] and Sirt6 [25] can catalyze the removal of long chain fatty acyl groups (e.g., myristoyl group) from lysine residues. Sirt5 was recently demonstrated to preferentially hydrolyze malonyl, succinyl and glutaryl lysine [26–29]. In this study, we detected the decrotonylation activity of Sirt3, which further expanding the landscape of PTMs that are regulated by sirtuins. It also links cellular metabolic modulation with chromatin epigenetic regulation and provides opportunity to explore mechanisms adopted by Sirt3 in cellular regulation.

Among Sirt1-3, which showed decrotonylation activity towards both lysine crotonylated histone peptides and proteins in vitro, only Sirt3 was demonstrated to be an endogenous 'eraser' to regulate dynamics of histone crotonylation in living cells. In contrast, the manipulation of endogenous Sirt1 and Sirt2 expression resulted in undetectable effects on neither histone global nor H3K4 crotonylation. However, we cannot rule out the possibility that these two sirtuins could function as decrotonylases at other histone crotonylation sites. In addition, the immunoblotting analysis using pan anti-KCr antibody showed that multiple proteins, in addition to histones, were modified by lysine crotonylation. The crotonylation level on those non-histone proteins was not affected by knockdown of neither Sirt1-2 nor Sirt3 by siRNA, indicating there exist other decrotonylases rather than sirtuin proteins in control of dynamics of lysine crotonylation on those proteins. Therefore, further studies are needed to systematically profile mammalian crotonylome, analyze the lysine crotonylation sites targeted by Sirt1-3 and identify decrotonylases for non-histone proteins.

Experimental Methods
Cell Culture
HeLa S3, HEK293T and HeLa cells were cultured in DMEM supplemented with 10% fetal bovine serum (FBS), 100 U/mL penicillin and 100 μg/mL streptomycin. Cells were maintained in a humidified 37 °C incubator with 5% CO_2.

Stable Isotope Labeling of Amino Acids in Cell Culture (SILAC) Cell Culture
HeLa S3 cells were grown in suspension at 37 °C in a humidified atmosphere with 5% CO_2 in DMEM medium (–Arg, –Lys, Invitrogen) containing 10% dialyzed FBS (Invitrogen), penicillin-streptomycin and supplemented with 22 mg/L $^{13}C_6^{15}N_4$-L-arginine (Cambridge Isotope) and 50 mg/L $^{13}C_6^{15}N_2$-L-lysine (Cambridge Isotope) or the corresponding non-labeled amino acids (PEPTIDE INTERNATIONAL), respectively. Harvested cell pellets were washed with ice-cold phosphate-buffered saline (PBS) and frozen with liquid N_2. The cell powder grinded with a Ball Mill (Retch MM301) was stored at −80 °C until use.

Preparation of Whole-Cell Lysates for CLASPI Experiment
To prepare whole-cell lysates, the frozen cell powder was first resuspended in a hypotonic buffer (10 mM HEPES, pH 7.5, 2 mM $MgCl_2$, 0.1% tween-20, 20% glycerol, 2 mM PMSF, and Roche Complete EDTA-free protease inhibitors) and incubated for 10 min at 4 °C. The suspension was centrifuged at 16,000 g for 15 min at 4 °C and the supernatant was kept for use later. The pellet was resuspended in a high-salt buffer (50 mM HEPES, pH 7.5, 420 mM NaCl, 2 mM $MgCl_2$, 0.1% tween-20, 20% glycerol, 2 mM PMSF, and Roche Complete EDTA-free protease inhibitors) and incubated for 30 min at 4 °C. The suspension was centrifuged at 16,000 g for 15 min at 4 °C and the supernatant was combined with the soluble fraction in hypotonic buffer to give the whole-cell lysates.

CLASPI-Photo-Cross-Linking
In a 'selectivity filter' experiment, probe **1** and probe **C** were incubated with heavy and

light SILAC whole-cell lysates, respectively, in the binding buffer (50 mM HEPES, pH 7.5, 168 mM NaCl, 2 mM $MgCl_2$, 0.1% tween-20, 20% glycerol, 2 mM PMSF, Roche Complete EDTA-free protease inhibitor cocktail) for 15 min at 4 °C. The samples were then irradiated at 365 nm using a Spectroline ML-3500S UV lamp for 15 min on ice. In 'an affinity filter experiment', the heavy and light SILAC lysates were reacted with probe **1** in the absence and presence, respectively, of H3K4Cr (1–15) peptide (30 μM) as a competitor. After photo-cross-linking, the heavy and light lysates were pooled.

Cu(I)-Catalyzed Cycloaddition/Click Chemistry
Briefly, to the prepared samples, 100 μM rhodamine-azide for in-gel fluorescence scanning or cleavable biotin azide for streptavidin enrichment was added, followed by 1 mM tris(2-carboxyethyl)phosphine (TCEP), 100 μM tris[(1-benzyl-1H-1,2,3-triazol-4-yl)methyl]amine (TBTA) and finally the reactions were initiated by the addition of 1 mM $CuSO_4$. The reactions were incubated for 1.5 h at room temperature.

Streptavidin Affinity Enrichment of Biotinylated Proteins
After the click chemistry with cleavable biotin azide, the reaction was quenched by adding 4 volumes of ice-cold acetone to precipitate proteins. After washing with ice-cold methanol twice, the air-dried protein pellet was dissolved in PBS with 4% SDS, 20 mM EDTA and 10% glycerol by vortexting and heating. The solution was then diluted with PBS to give a final concentration of SDS of 0.5%. High-capacity streptavidin agarose beads (Thermo Pierce) were added to bind the biotinylated proteins with rotating for 1.5 h at room temperature. To remove nonspecific binding, the beads were washed with PBS with 0.2% SDS, 6 M urea in PBS with 0.1% SDS and 250 mM NH_4HCO_3 with 0.05% SDS. The enriched proteins were then eluted by incubating with 25 mM $Na_2S_2O_4$, 250 mM NH_4HCO_3 and 0.05% SDS for 1 h. The eluted proteins were dried down with SpeedVac.

Sample Preparation for Mass Spectrometry
The dried proteins were resuspended in 30 μL LDS sample loading buffer (Invitrogen) with 50 mM dithiothreitol (DTT) and heated at 75 °C for 8 min, then reacted with iodoacetamide in dark for 30 min to alkylate all reduced cysteines. Proteins were then separated on a Bis-Tris gel, followed by fixation in a 50% methanol/7% acetic acid solution. The gel was stained by GelCode Blue stain (Pierce). The diced 1 mm^3 cubes of gels were then destained by incubating with 50 mM ammonium bicarbonate/50% acetonitrile for 1 h. The destained gel cubes were dehydrated in acetonitrile for 10 min and rehydrated in 25 mM NH_4HCO_3 with trypsin for protein digestion at 37 °C overnight. The resulting peptides were enriched with the StageTips. The peptides eluted from the StageTips were dried down by SpeedVac and then resuspended in 0.5% acetic acid for analysis by LC-MS/MS.

Mass Spectrometry
Mass spectrometry was performed on an LTQ-Orbitrap Velos mass spectrometer (Thermo Fisher Scientific). First, peptide samples in 0.1% formic acid were pressure

loaded onto a self-packed PicoTip column (New Objective) (360-μm o.d., 75-μm i.d., 15-μm tip), packed with 7–10 cm of reverse-phase C18 material (ODS-A C18 5-μm beads from YMC), rinsed for 5 min with 0.1% formic acid and subsequently eluted with a linear gradient from 2 to 35% B in 150 min (A = 0.1% formic acid, B = 0.1% formic acid in ACN, flow rate ~200 nL/min) into the mass spectrometer. The instrument was operated in a data dependent mode cycling through a full scan (300–2,000 m/z, single μscan) followed by 10 CID MS/MS scans on the 10 most abundant ions from the immediate preceding full scan. The cations were isolated with a 2-Da mass window and set on a dynamic exclusion list for 60 s after they were first selected for MS/MS. The raw data were processed and analyzed using MaxQuant (version 1.2.2.5). A human fasta file (ipi.HUMAN.v.3.68.fasta) was used as protein sequence searching database. Default parameters were adapted for the protein identification and quantification. In particular, parent peak MS tolerance is 6 ppm, MS/MS tolerance is 0.5 Da, the minimum peptide length is 6 amino acids, the maximum number of missed cleavages is 2. The proteins quantified were supported by at least 2 quantification events. Both the 'selectivity-filter' and 'affinity-filter' experiments were repeated twice, and only the proteins that were identified and quantified in all experiments were reported.

In-gel Fluorescence Visualization

The click chemistry reactions were quenched by adding 1 volume of 2 × sample buffer. The proteins were heated at 85 °C for 8 min, and resolved by SDS-PAGE. The labeled proteins were visualized by scanning the gel on a Typhoon 9410 variable mode imager (excitation 532 nm, emission 580 nm).

Expression and Purification of Recombinant Human Sirtuins

Plasmids of Sirt1 (193–747), Sirt2 (36–356), Sirt5 (34–302) and Sirt6 (1–314) for *E. coli* expression were generated as previously described. Plasmids of Sirt3 (102–399) cloned in pTrcHis 2C vector for *E.coli* expression and full length Sirt3 (wide-type and mutant H248Y) cloned into pcDNA3.1 vector for mammalian cell expression were generous gifts from Eric Verdin (UCSF). Sirt3 mutant F180A was generated by site-directed mutagenesis. All the proteins were expressed in *E.coli* Rosetta cells. To induce the expression of target proteins, isopropyl β-D-1-thiogalactopyranoside (IPTG) was added to a final concentration of 0.2 mM when OD_{600} reached 0.6, and the culture was grown at 15 °C (Sirt3 at 25 °C) for 16–18 h. Cells were harvested and resuspended in lysis buffer A (50 mM Tris-HCl, pH 7.5, 500 mM NaCl, 1 mM PMSF and Roche EDTA-free protease inhibitors, for Sirt1, Sirt2 and Sirt6) or buffer B (50 mM Tris-HCl, pH 7.5, 150 mM NaCl, 1 mM PMSF and Roche EDTA-free protease inhibitors, for Sirt3 and Sirt5). Following sonication and centrifugation, the supernatant was loaded onto a nickel column pre-equilibrated with lysis buffer. The column was washed with 5 column volumes of wash buffer (lysis buffer with 30 mM Imidazole) and then the target proteins were eluted with elution buffer (lysis buffer with 250 mM Imidazole). After purification, Sirt2 was digested by UPL1 at 4 °C for overnight and purified by a Highload 26/60 Superdex75 gel-filtration column (GE

Healthcare Life Sciences). Sirt6 was purified by SP-column and Superdex 75 gel-filtration column. Others were loaded onto a Superdex75 gel-filtration or Highload 26/60 Superdex200 (for Sirt1) column. After concentration, the target proteins were frozen and then stored at −80 °C.

Isothermal Titration Calorimetry Measurements

Experiments were performed at 25 °C on a MicroCal iTC200 titration calorimeter (MicroCal). The reaction cell containing 200 μL of 100–200 μM proteins was titrated with 17 injections, first 0.5 μL, and all subsequent injections of 2 μL of 1.5–2.5 mM peptides. The binding isotherm was fit with Origin 7.0 software package (OriginLab) that uses a single set of independent sites to determine the thermodynamic binding constants and stoichiometry.

Enzymatic Reactions

The enzymatic activities of human Sirtuins were measured by detecting the removal of crotonyl group from peptides [26]. 5 μM of each Sirtuin protein was incubated with 500 μM of corresponding crotonylated peptides and 1 mM of nicotinamide adenine dinucleotide (NAD) in a reaction buffer containing 20 mM Tris-HCl buffer (pH 7.5) and 1 mM DTT at 37 °C for 2 h. The reactions were stopped by adding 1/3 reaction volume of 20% TFA and frozen in liquid N_2 immediately. For Sirt3, samples without NAD or without enzyme were treated with the same condition as controls. Samples were then analyzed by LC-MS with a Vydac 218TP C18 column (4.6 mm × 250 mm, 5 μm, Grace Davison). Mobile phases used were 0.05% TFA in water (buffer A) and 0.05% TFA in ACN (buffer B). The flow rate for LC was 0.6 mL/min. The peptide mixtures were eluted by buffer A for 10 min, then 0–30% buffer B over 10 min. MS started to record at 10 min for each injection.

RNAi Experiments

15 nM Sirt1 siRNA (Santa Cruz), 30 nM Sirt2 siRNA (Thermo Scientific) or 30 nM Sirt3 siRNA (Thermo Scientific) were transfected into HeLa cell line with Dharma-FECT 1 Transfection Reagent (Thermo Scientific), according to the manufacturer's instructions. Corresponding concentration of control siRNA was used as negative control. Following transfection, cells were then maintained in a humidified 37 °C incubator with 5% CO_2 for another 48 h (for Sirt1 and Sirt2) or 72 h (for Sirt3).

Histone Extraction

Acid-extraction method was used to isolate histones from HeLa S3 cells [30]. Briefly, the harvested HeLa S3 cell pellet was resuspended with lysis buffer (10 mM Tris–HCl pH 8.0, 1 mM KCl, 1.5 mM $MgCl_2$ and 1 mM DTT 2 mM PMSF and Roche Complete EDTA-free protease inhibitors) and incubated at 4 °C by rotating for 1 h. The intact nuclei were pelleted by centrifuging at 10000 g for 10 min at 4 °C. To extract histones, 0.4 N H_2SO_4 was added to resuspend the nuclei, followed by rotating at 4 °C for overnight. After centrifuging to remove the nuclei debris, histones were precipitated by adding 100% trichloroacetic acid (TCA) drop by drop (TCA final concentration: 33%). The precipitated histones were pelleted at 16,000 g for 10 min at 4 °C and washed with ice-cold acetone twice. The air-dried protein pellet was dissolved with ddH_2O and stored at −80 °C for later use.

On-membrane Decrotonylation Experiment

20 μg of HeLa S3 whole-cell lysate or 5 μg extracted histones were resolved by SDS-PAGE gel and transferred to the poly(vinylidene fluoride) (PVDF) membranes. The membranes were incubated with or without 0.1 μM of Sirt3 in reaction buffer (25 mM Tris-HCl, 130 mM NaCl, 3 mM KCl, 1 mM $MgCl_2$, 1 mM DTT, pH 7.5) containing 1 mM NAD at 37 °C for 2 h.

Immunoblotting

Proteins separated by SDS-PAGE were transferred onto a PVDF membrane which was then blocked (5% nonfat dried milk and 0.1% tween-20 in PBS) for 1 h at room temperature. The membrane was incubated with primary antibody diluted in PBST (0.1% Tween-20 in PBS) with 2% BSA, followed by washing with PBST for 5 min trice, incubated with goat anti-rabbit-HRP conjugated secondary antibody (1:20000, Santa Cruz), or rabbit anti-mouse-HRP conjugated secondary antibody (1:5000, Santa Cruz) diluted in PBST for 1 h at room temperature, and then visualized with western blotting detection reagents (Thermo).

Subcellular Fractionation

In brief, HeLa cells were harvested by centrifugation and washed with PBS twice, subsequent all steps were performed at 4 °C. Then the cells were suspended by 5 cell pellet volumes of buffer A (10 mM HEPES, pH 7.9 at 4 degrees, 1.5 mM $MgCl_2$, 10 mM KC1 and 0.5 mM DTT) following by incubation for 10 min. After centrifugation, cells were re-suspended in 2 cell pellet volumes of buffer A and lysed by Dounce homogenizer (B type pestle) with homogenate checked by microscopy. The cell lysis was layered over 30% sucrose in buffer A and then centrifuge for 15 min at 800 g. The resulting pellet was recovered from sucrose phase and wash by buffer A twice then extracted by buffer C (20 mM HEPES, pH 7.9, 25% (v/v) glycerol, 0.42 M NaCl, 1.5 mM $MgCl_2$, 0.2 mM EDTA, 0.5 mM PMSF and 0.5 mM DTT) for 30 min at 4 °C. After centrifuge at 12, 000 g for 30 min, the supernatant was termed as nuclear fraction. The resulting supernatant from 800 g centrifuge was centrifuged twice at 800 g to complete pellet nuclei or intact cell. Then the supernatant was centrifuged at 7, 000 g to pellet mitochondria followed by wash twice by buffer A. Then the mitochondria were lysed by TXIP-1 buffer (1% Triton X-100 [v/v], 150 mM NaCl, 0.5 mM EDTA, 50 mM Tris-HCl, pH 7.4). Protein concentration was determined by BCA assay.

References

1. Kouzarides T (2007) Chromatin modifications and their function. Cell 128:693–705. https://doi.org/10.1016/j.cell.2007.02.005

2. Jenuwein T, Allis CD (2001) Translating the histone code. Science 293:1074–1080. https://doi.org/10.1126/science.1063127
3. Goldberg AD, Allis CD, Bernstein E (2007) Epigenetics: a landscape takes shape. Cell 128:635–638. https://doi.org/10.1016/j.cell.2007.02.006
4. Seet BT, Dikic I, Zhou MM, Pawson T (2006) Reading protein modifications with interaction domains. Nat Rev Mol Cell Biol 7:473–483. https://doi.org/10.1038/nrm1960
5. Taverna SD, Li H, Ruthenburg AJ, Allis CD, Patel DJ (2007) How chromatin-binding modules interpret histone modifications: lessons from professional pocket pickers. Nat Struct Mol Biol 14:1025–1040. https://doi.org/10.1038/nsmb1338
6. Tan M et al (2011) Identification of 67 histone marks and histone lysine crotonylation as a new type of histone modification. Cell 146:1016–1028. https://doi.org/10.1016/j.cell.2011.08.008
7. Montellier E, Rousseaux S, Zhao Y, Khochbin S (2012) Histone crotonylation specifically marks the haploid male germ cell gene expression program: post-meiotic male-specific gene expression. BioEssays News Rev Mol Cell Dev Biol 34:187–193. https://doi.org/10.1002/bies.201100141
8. Madsen AS, Olsen CA (2012) Profiling of substrates for zinc-dependent lysine deacylase enzymes: HDAC3 exhibits decrotonylase activity in vitro. Angew Chem 51:9083–9087. https://doi.org/10.1002/anie.201203754
9. Feldman JL, Baeza J, Denu JM (2013) Activation of the protein deacetylase SIRT6 by long-chain fatty acids and widespread deacylation by mammalian sirtuins. J Biol Chem 288:31350–31356. https://doi.org/10.1074/jbc.c113.511261
10. Li X, Foley EA, Molloy KR, Li Y, Chait BT, Kapoor TM (2012) Quantitative chemical proteomics approach to identify post-translational modification-mediated protein-protein interactions. J Am Chem Soc 134:1982–1985. https://doi.org/10.1021/ja210528v
11. Li X, Kapoor TM (2010) Approach to profile proteins that recognize post-translationally modified histone "tails". J Am Chem Soc 132:2504–2505. https://doi.org/10.1021/ja909741q
12. Imai S, Armstrong CM, Kaeberlein M, Guarente L (2000) Transcriptional silencing and longevity protein Sir2 is an NAD-dependent histone deacetylase. Nature 403:795–800. https://doi.org/10.1038/35001622
13. Landry J, Sutton A, Tafrov ST, Heller RC, Stebbins J, Pillus L, Sternglanz R (2000) The silencing protein SIR2 and its homologs are NAD-dependent protein deacetylases. Proc Nal Acad Sci U S A 97:5807–5811. https://doi.org/10.1073/pnas.110148297
14. Sauve AA, Wolberger C, Schramm VL, Boeke JD (2006) The biochemistry of sirtuins. Annu Rev Biochem 75:435–465. https://doi.org/10.1146/annurev.biochem.74.082803.133500
15. Tanner KG, Landry J, Sternglanz R, Denu JM (2000) Silent information regulator 2 family of NAD- dependent histone/protein deacetylases generates a unique product, 1-O-acetyl-ADP-ribose. Proc Natl Acad Sci U S A 97:14178–14182. https://doi.org/10.1073/pnas.250422697
16. Jin L et al (2009) Crystal structures of human SIRT3 displaying substrate-induced conformational changes. J Biol Chem 284:24394–24405. https://doi.org/10.1074/jbc.m109.014928
17. Hirschey MD, Shimazu T, Huang JY, Schwer B, Verdin E (2011) SIRT3 regulates mitochondrial protein acetylation and intermediary metabolism. Cold Spring Harb Symp Quant Biol 76:267–277. https://doi.org/10.1101/sqb.2011.76.010850
18. Shi T, Wang F, Stieren E, Tong Q (2005) SIRT3, a mitochondrial sirtuin deacetylase, regulates mitochondrial function and thermogenesis in brown adipocytes. J Biol Chem 280:13560–13567. https://doi.org/10.1074/jbc.m414670200
19. Hirschey MD et al (2011) SIRT3 deficiency and mitochondrial protein hyperacetylation accelerate the development of the metabolic syndrome. Mol Cell 44:177–190. https://doi.org/10.1016/j.molcel.2011.07.019
20. Scher MB, Vaquero A, Reinberg D (2007) SirT3 is a nuclear NAD + -dependent histone deacetylase that translocates to the mitochondria upon cellular stress. Genes Dev 21:920–928. https://doi.org/10.1101/gad.1527307
21. Iwahara T, Bonasio R, Narendra V, Reinberg D (2012) SIRT3 functions in the nucleus in the control of stress-related gene expression. Mol Cell Biol 32:5022–5034. https://doi.org/10.1128/mcb.00822-12

22. Bao X et al. (2014) Identification of 'erasers' for lysine crotonylated histone marks using a chemical proteomics approach. eLife 3. https://doi.org/10.7554/elife.02999
23. Liu Z, Yang T, Li X, Peng T, Hang HC, Li XD (2015) Integrative chemical biology approaches for identification and characterization of "erasers" for fatty-acid-acylated lysine residues within proteins. Angew Chem 54:1149–1152. https://doi.org/10.1002/anie.201408763
24. Teng YB et al (2015) Efficient demyristoylase activity of SIRT2 revealed by kinetic and structural studies. Sci Rep 5:8529. https://doi.org/10.1038/srep08529
25. Jiang H et al (2013) SIRT6 regulates TNF-alpha secretion through hydrolysis of long-chain fatty acyl lysine. Nature 496:110–113. https://doi.org/10.1038/nature12038
26. Du J et al (2011) Sirt5 is a NAD-dependent protein lysine demalonylase and desuccinylase. Science 334:806–809. https://doi.org/10.1126/science.1207861
27. Rardin MJ et al (2013) SIRT5 regulates the mitochondrial lysine succinylome and metabolic networks. Cell Metab 18:920–933. https://doi.org/10.1016/j.cmet.2013.11.013
28. Peng C et al (2011) The first identification of Lysine Malonylation substrates and its regulatory enzyme. Mol cell Proteomics MCP 10(12):M111.012658. https://doi.org/10.1074/mcp.M111.012658
29. Tan M et al (2014) Lysine glutarylation is a protein posttranslational modification regulated by SIRT5. Cell Metab 19:605–617. https://doi.org/10.1016/j.cmet.2014.03.014
30. Shechter D, Dormann HL, Allis CD, Hake SB (2007) Extraction, purification and analysis of histones. Nat Protoc 2:1445–1457. https://doi.org/10.1038/nprot.2007.202

Appendix

See Table A.1.

Table A.1 The malonylated protein candidates identified using MalAM-yne

No.	Protein description	Protein ID	Total spectral counts			
			Expt #1		Expt #2	
			Control	Probe	Control	Probe
1	14-3-3 protein theta	IPI00018146	43	99	26	56
2	2,4-dienoyl-CoA reductase, mitochondrial	IPI00003482	0	9	0	5
3	Transgelin-2	IPI00647915	62	144	17	57
4	GTP-binding nuclear protein Ran	IPI00792352	16	54	12	29
5	26S proteasome non-ATPase regulatory subunit 12	IPI00185374	0	8	0	7
6	26S proteasome non-ATPase regulatory subunit 2	IPI00012268	8	87	0	8
7	40S ribosomal protein SA	IPI00413108	10	23	0	6
8	40S ribosomal protein S12	IPI00013917	15	40	2	17
9	40S ribosomal protein S17	IPI00221093	10	39	0	5
10	40S ribosomal protein S3	IPI00011253	5	28	2	9
11	40S ribosomal protein S5	IPI00008433	22	179	5	29
12	40S ribosomal protein S7	IPI00013415	0	6	0	8
13	40S ribosomal protein S8	IPI00216587	0	10	2	6
14	40S ribosomal protein S9	IPI00221088	2	10	0	8
15	60S ribosomal protein L10a	IPI00412579	8	18	2	10
16	6-phosphogluconolactonase	IPI00029997	9	49	3	16
17	Abhydrolase domain-containing protein 10, mitochondrial	IPI00020075	0	10	0	5
18	Acetyl-CoA acetyltransferase, mitochondrial	IPI00030363	2	15	0	10

(continued)

X. Bao, *Study on the Cellular Regulation and Function of Lysine Malonylation, Glutarylation and Crotonylation*, Springer Theses,
https://doi.org/10.1007/978-981-15-2509-4

Table A.1 (continued)

No.	Protein description	Protein ID	Total spectral counts			
			Expt #1		Expt #2	
			Control	Probe	Control	Probe
19	Adenylate kinase 1	IPI00640817	5	22	1	7
20	ADP-ribosylation factor 4	IPI00215918	2	12	0	6
21	Aldo-keto reductase family 1 member C3	IPI00291483	10	34	5	10
22	Aldose reductase	IPI00413641	11	39	0	9
23	Annexin A1	IPI00218918	38	77	16	43
24	Annexin A3	IPI00024095	6	23	4	14
25	Annexin A5	IPI00329801	30	68	5	29
26	Aspartate aminotransferase, mitochondrial	IPI00018206	14	47	6	23
27	ATP synthase subunit alpha, mitochondrial	IPI00440493	4	41	4	44
28	ATP synthase subunit b, mitochondrial	IPI00029133	13	31	3	16
29	ATP synthase subunit beta, mitochondrial	IPI00303476	6	41	5	48
30	ATP synthase subunit O, mitochondrial	IPI00007611	6	19	1	10
31	X-ray repair cross-complementing protein 6	IPI00644712	12	50	3	12
32	ATP-dependent DNA helicase 2 subunit 2	IPI00220834	29	189	5	62
33	ATP-dependent RNA helicase A	IPI00844578	74	151	7	66
34	ATP-dependent RNA helicase DDX3X	IPI00215637	16	85	4	51
35	BAG family molecular chaperone regulator 2	IPI00000643	0	24	0	8
36	Basic leucine zipper and W2 domain-containing protein 2	IPI00022305	2	10	2	5
37	B-cell receptor-associated protein 31 isoform a	IPI00642984	0	18	2	5
38	Bifunctional ATP-dependent dihydroxyacetone kinase/FAD-AMP lyase (cyclizing)	IPI00551024	0	10	0	5
39	Bifunctional purine biosynthesis protein PURH	IPI00289499	11	57	2	14
40	CAD protein	IPI00301263	14	48	3	13
41	Calpain small subunit 1	IPI00025084	4	51	0	12
42	Carbamoyl-phosphate synthetase 1 isoform a precursor, mitochondrial	IPI00889534	230	630	63	348
43	Carboxymethylenebutenolidase homolog	IPI00383046	2	43	0	11
44	Serine-threonine kinase receptor-associated protein	IPI00294536	16	33	2	5
45	Calponin-2	IPI00910593	0	21	2	8
46	Karyopherin alpha 6	IPI00747764	7	19	0	8

(continued)

Table A.1 (continued)

No.	Protein description	Protein ID	Total spectral counts			
			Expt #1		Expt #2	
			Control	Probe	Control	Probe
47	Isoform 1 of Hematological and neurological expressed 1-like protein	IPI00909195	0	19	0	10
48	Eukaryotic translation initiation factor 4B	IPI00012079	9	29	5	27
49	Cell division control protein 42 homolog	IPI00909484	5	22	3	8
50	Transketolase	IPI00643920	30	152	18	66
51	Calnexin	IPI00020984	15	45	4	23
52	Homo sapiens transforming growth factor beta regulator 4 (TBRG4)	IPI00329625	0	18	0	7
53	Isoform 2 of Very long-chain specific acyl-CoA dehydrogenase, mitochondrial	IPI00028031	3	49	1	26
54	T-lymphokine-activated killer cell-originated protein kinase	IPI00306708	2	22	0	7
55	Adenylosuccinate lyase	IPI00026904	1	25	0	6
56	Eukaryotic translation initiation factor 4gamma 2	IPI00015952	3	15	0	5
57	Mitochondrial dicarboxylate carrier	IPI00005537	2	28	0	14
58	Alanyl-tRNA synthetase, cytoplasmic	IPI00910701	53	127	6	17
59	Succinate dehydrogenase (ubiquinone) flavoprotein subunit, mitochondrial	IPI00305166	3	35	0	8
60	Isoform 2 of Clathrin interactor 1	IPI00291930	0	5	0	6
61	Glutaminyl-tRNA synthetase	IPI00026665	9	67	1	9
62	Aminoacyl tRNA synthetase complex-interacting multifunctional protein 2	IPI00940113	1	10	0	5
63	Citrate synthase, mitochondrial	IPI00025366	35	94	8	18
64	Cleavage and polyadenylation specificity factor subunit 5	IPI00646917	0	19	0	8
65	Coatomer subunit zeta-1	IPI00032851	0	21	0	5
66	Complement component 1 Q subcomponent-binding protein, mitochondrial	IPI00014230	4	21	2	15
67	Condensin complex subunit 1	IPI00299524	5	19	0	7
68	Copine-3	IPI00024403	0	19	0	6
69	Coronin-1C	IPI00867509	3	38	0	7
70	CTP synthase 1	IPI00290142	3	40	0	5
71	Cystatin-B	IPI00021828	5	26	6	12

(continued)

Table A.1 (continued)

No.	Protein description	Protein ID	Total spectral counts			
			Expt #1		Expt #2	
			Control	Probe	Control	Probe
72	Cysteine and glycine-rich protein 1	IPI00442073	4	16	0	10
73	Cytochrome b-c1 complex subunit 2, mitochondrial	IPI00305383	9	49	3	7
74	D-3-phosphoglycerate dehydrogenase; Phosphoglycerate dehydrogenase	IPI00011200	15	73	3	36
75	DEAD box polypeptide 17 isoform 3	IPI00940000	1	28	6	25
76	Deoxyribonucleoside 5-monophosphate N-glycosidase; putative c-Myc-responsive isoform 2	IPI00007926	2	29	0	5
77	DNA damage-binding protein 1	IPI00293464	31	82	1	23
78	DNA-(apurinic or apyrimidinic site) lyase	IPI00215911	7	45	1	9
79	DnaJ homolog subfamily A member 1	IPI00012535	5	27	2	12
80	Dolichyl-diphosphooligosaccharide–protein glycosyltransferase subunit 1 precursor	IPI00025874	8	32	2	13
81	Dynactin 1 isoform 3	IPI00935906	0	16	0	14
82	EIF4G1 protein	IPI00925413	51	125	3	120
83	Elongation factor 1-alpha 1	IPI00396485	232	661	56	261
84	Elongation factor 1-gamma	IPI00937615	125	251	38	88
85	Elongation factor 2	IPI00186290	217	743	58	254
86	Emerin	IPI00032003	1	6	0	12
87	Endoplasmic reticulum protein ERp29	IPI00024911	3	39	0	9
88	Enoyl-CoA hydratase, mitochondrial	IPI00024993	2	16	0	19
89	Eukaryotic peptide chain release factor subunit 1	IPI00429191	2	14	0	6
90	Eukaryotic translation initiation factor 2 subunit 3	IPI00297982	3	56	8	23
91	Eukaryotic translation initiation factor 3 subunit A	IPI00029012	2	25	0	24
92	Eukaryotic translation initiation factor 3 subunit M	IPI00102069	9	50	6	24
93	Ewing sarcoma breakpoint region 1 isoform 1	IPI00009841	1	8	0	5
94	Ezrin	IPI00843975	8	69	2	15
95	FACT complex subunit SPT16	IPI00026970	42	85	3	35
96	Fatty acid synthase	IPI00026781	145	336	34	139
97	FK506-binding protein 3	IPI00024157	0	6	0	6
98	FK506-binding protein 4	IPI00219005	7	34	4	9

(continued)

Table A.1 (continued)

No.	Protein description	Protein ID	Total spectral counts			
			Expt #1		Expt #2	
			Control	Probe	Control	Probe
99	Flap endonuclease 1	IPI00026215	8	24	2	9
100	Flavin reductase	IPI00783862	14	46	1	12
101	FSCN1 protein (Fragment); Fascin	IPI00747810	15	75	1	6
102	Fus-like protein (Fragment)	IPI00260715	6	39	2	30
103	G1 to S phase transition 1 isoform 2	IPI00909083	1	27	0	7
104	Galectin-1	IPI00219219	5	18	0	7
105	Galectin-3	IPI00465431	11	27	4	19
106	Gem-associated protein 5	IPI00291783	3	14	0	9
107	Glutaredoxin-3	IPI00008552	6	20	2	10
108	Glutathione S-transferase kappa 1 isoform b	IPI00440703	4	28	0	5
109	Glutathione S-transferase Mu 3	IPI00246975	4	51	0	11
110	Glutathione S-transferase omega-1	IPI00019755	4	64	1	30
111	GMP synthase [glutamine-hydrolyzing]	IPI00029079	7	53	2	17
112	GrpE protein homolog 1	IPI00029557	0	19	0	13
113	GTP-binding protein SAR1a	IPI00015954	12	35	0	6
114	Guanine aminohydrolase	IPI00873506	3	16	3	11
115	Eukaryotic translation initiation factor 3 subunit F	IPI00654777	4	30	3	9
116	Heat shock 70 kDa protein 1	IPI00304925	93	259	51	181
117	Heat shock 70 kDa protein 4	IPI00002966	41	116	0	34
118	Heat shock protein 75 kDa, mitochondrial	IPI00030275	7	61	0	29
119	Heterogeneous nuclear ribonucleoprotein L	IPI00027834	19	76	4	23
120	HLA-B associated transcript 3	IPI00892541	5	12	0	6
121	Hsc70-interacting protein	IPI00032826	1	13	1	5
122	Hsp90 co-chaperone Cdc37	IPI00013122	6	23	3	8
123	HSPA5 protein	IPI00003362	20	71	20	48
124	Niban-like protein 1	IPI00647286	1	26	1	17
125	Hypoxia up-regulated protein 1	IPI00000877	7	21	0	20
126	Karyopherin alpha 2	IPI00002214	10	72	0	15
127	Importin subunit beta-1	IPI00001639	76	206	17	36
128	Importin-9	IPI00185146	37	74	2	28
129	182 kDa tankyrase-1-binding protein	IPI00304589	0	13	0	11
130	Hydroxysteroid (17-beta) dehydrogenase 10	IPI00017726	26	89	3	41
131	Ribosomal protein L11	IPI00376798	2	28	2	6

(continued)

Table A.1 (continued)

No.	Protein description	Protein ID	Total spectral counts			
			Expt #1		Expt #2	
			Control	Probe	Control	Probe
132	60S ribosomal protein L12	IPI00024933	13	27	0	9
133	Acidic (Leucine-rich) nuclear phosphoprotein 32 family, member B	IPI00007423	3	13	1	5
134	Adenylate kinase 2, mitochondrial	IPI00215901	1	29	0	14
135	Annexin A7	IPI00002460	4	16	1	13
136	ATP synthase subunit d, mitochondrial	IPI00220487	15	32	0	22
137	ATP-dependent RNA helicase DDX42	IPI00409671	2	7	0	10
138	Caprin-1	IPI00783872	9	28	1	23
139	Cell division cycle and apoptosis regulator protein 1	IPI00217357	0	7	0	13
140	Cysteine and histidine-rich domain-containing protein 1	IPI00015897	0	17	0	6
141	Disabled homolog 2	IPI00179438	0	22	0	13
142	DNA (cytosine-5)-methyltransferase 1	IPI00031519	7	16	1	5
143	DNA-dependent protein kinase catalytic subunit	IPI00296337	64	151	14	89
144	Exportin-5	IPI00640703	16	43	0	12
145	Far upstream element-binding protein 3	IPI00377261	1	22	0	6
146	Gamma-glutamylcyclotransferase	IPI00031564	7	15	0	6
147	Isoform 1 of General transcription factor II-I	IPI00054042	17	58	2	46
148	Isoform 1 of Heat shock cognate 71 kDa protein	IPI00003865	76	220	32	105
149	Isoform 1 of Helicase-like transcription factor	IPI00339381	0	9	0	6
150	Isoform 1 of Heterogeneous nuclear ribonucleoprotein Q	IPI00018140	5	48	1	10
151	Isoform 1 of Hydroxysteroid dehydrogenase-like protein 2	IPI00414384	0	5	0	6
152	Isoform 1 of Kinectin	IPI00328753	2	6	0	6
153	Isoform 1 of La-related protein 1	IPI00185919	2	17	2	12
154	Isoform 1 of Leukotriene A-4 hydrolase	IPI00219077	18	70	6	15
155	Isoform 1 of Microtubule-associated protein 4	IPI00396171	10	68	0	65
156	Isoform 1 of Myoferlin	IPI00021048	21	46	0	7

(continued)

Table A.1 (continued)

No.	Protein description	Protein ID	Total spectral counts			
			Expt #1		Expt #2	
			Control	Probe	Control	Probe
157	Isoform 1 of Myopalladin	IPI00645179	2	27	0	15
158	Isoform 1 of Nodal modulator 2	IPI00465432	5	25	0	8
159	Isoform 1 of Nuclear pore complex protein Nup155	IPI00026625	7	48	0	26
160	Isoform 1 of Nucleophosmin	IPI00549248	33	66	7	42
161	Isoform 1 of Plasminogen activator inhibitor 1 RNA-binding protein	IPI00410693	2	21	5	19
162	Isoform 1 of Poly(U)-binding-splicing factor PUF60	IPI00069750	4	43	0	16
163	Isoform 1 of Polyadenylate-binding protein 1	IPI00008524	19	65	12	59
164	Isoform 1 of Proteasome subunit alpha type-3; Isoform 2 of Proteasome subunit alpha type-3	IPI00419249	1	25	1	5
165	Isoform 1 of Proteasome subunit alpha type-7	IPI00024175	3	20	1	16
166	Isoform 1 of Protein diaphanous homolog 1	IPI00852685	9	31	0	26
167	Isoform 1 of Protein KIAA1967	IPI00182757	3	26	0	14
168	Isoform 1 of Protein SET	IPI00072377	0	12	0	13
169	Isoform 1 of Protein transport protein Sec24A	IPI00873472	0	10	0	5
170	Isoform 1 of Protein-L-isoaspartate(D-aspartate) O-methyltransferase	IPI00411680	1	16	0	7
171	Isoform 1 of Regulation of nuclear pre-mRNA domain-containing protein 2	IPI00384541	0	7	0	6
172	Isoform 1 of Regulator of nonsense transcripts 1	IPI00034049	10	37	0	17
173	Isoform 1 of Ribosome-recycling factor	IPI00061108	0	12	0	12
174	Isoform 1 of RNA-binding protein 25	IPI00004273	1	6	0	6
175	Isoform 1 of RuvB-like 1	IPI00021187	2	24	1	5
176	Isoform 1 of Scavenger receptor class B member 1	IPI00177968	1	52	0	38
177	Isoform 1 of Signal transducer and activator of transcription 3	IPI00784414	4	23	0	5
178	Isoform 1 of Sister chromatid cohesion protein PDS5 homolog A	IPI00854642	3	17	0	9
179	Isoform 1 of S-phase kinase-associated protein 1	IPI00301364	1	22	0	5
180	Isoform 1 of Splicing factor 3B subunit 3	IPI00300371	24	70	0	18

(continued)

Table A.1 (continued)

No.	Protein description	Protein ID	Total spectral counts			
			Expt #1		Expt #2	
			Control	Probe	Control	Probe
181	Isoform 1 of Splicing factor U2AF 65 kDa subunit	IPI00031556	6	19	0	7
182	Isoform 1 of STE20-like serine/threonine-protein kinase	IPI00022827	2	10	0	6
183	Isoform 1 of Structural maintenance of chromosomes protein 2	IPI00007927	5	18	0	11
184	Isoform 1 of Structural maintenance of chromosomes protein 4	IPI00411559	7	24	0	10
185	Isoform 1 of Testisin	IPI00305703	0	9	0	8
186	Isoform 1 of Thioredoxin reductase 1	IPI00884896	32	135	2	27
187	Isoform 1 of Transcription elongation regulator 1	IPI00247871	0	7	0	8
188	Isoform 1 of Transcription factor BTF3	IPI00221035	0	11	0	9
189	Isoform 1 of U5 small nuclear ribonucleoprotein 200 kDa helicase	IPI00420014	18	40	0	14
190	Isoform 1 of Ubiquitin conjugation factor E4 B	IPI00005715	2	13	0	5
191	Isoform 1 of UPF0557 protein C10orf119	IPI00478758	0	52	4	20
192	Isoform 1 of YTH domain family protein 2	IPI00306043	1	14	1	6
193	Isoform 1 of Zinc finger CCCH-type antiviral protein 1	IPI00410067	1	6	0	6
194	Isoform 2 of AP-2 complex subunit beta	IPI00784366	18	60	0	8
195	Isoform 2 of Coatomer subunit alpha	IPI00646493	19	73	0	23
196	Isoform 2 of Eukaryotic translation initiation factor 3 subunit B	IPI00719752	21	79	0	19
197	Isoform 2 of Eukaryotic translation initiation factor 5A-1	IPI00376005	31	78	3	32
198	Isoform 2 of Extended synaptotagmin-1	IPI00746655	10	45	1	16
199	Isoform 2 of Heat shock protein HSP 90-alpha	IPI00382470	59	157	23	102
200	Isoform 2 of Hematological and neurological expressed 1 protein	IPI00384857	0	6	0	6
201	Isoform 2 of Hydroxyacyl-coenzyme A dehydrogenase, mitochondrial	IPI00298406	9	74	2	28
202	Isoform 2 of Myb-binding protein 1A	IPI00607584	1	8	0	8
203	Isoform 2 of PERQ amino acid-rich with GYF domain-containing protein 2	IPI00647635	0	11	0	6
204	Isoform 2 of Protein disulfide-isomerase A6	IPI00299571	5	20	6	18
205	Isoform 2 of Putative ATP-dependent RNA helicase DHX30	IPI00926109	0	15	0	7
206	Isoform 2 of Tropomyosin alpha-3 chain	IPI00218319	4	31	1	7

(continued)

Table A.1 (continued)

No.	Protein description	Protein ID	Total spectral counts			
			Expt #1		Expt #2	
			Control	Probe	Control	Probe
207	Isoform 2 of Ubiquitin-associated protein 2-like	IPI00029019	3	65	0	45
208	Isoform 3 of LIM domain only protein 7	IPI00291802	5	30	0	39
209	Isoform 3 of Nuclear autoantigenic sperm protein	IPI00332499	10	75	0	20
210	Isoform 3 of Tumor protein D52	IPI00873344	0	23	0	8
211	Isoform 4 of DNA topoisomerase 2-alpha	IPI00879004	15	39	0	20
212	Isoform 5 of Nuclear pore complex protein Nup214	IPI00646361	0	21	0	15
213	Isoform 5 of Serine/threonine-protein phosphatase 6 regulatory subunit 3	IPI00719725	4	24	0	6
214	Isoform 5 of Splicing factor 1	IPI00386119	0	19	0	6
215	Isoform 6 of GTPase-activating protein and VPS9 domain-containing protein 1	IPI00292753	1	29	0	22
216	Isoform 7 of BAT2 domain-containing protein 1'	IPI00083708	0	23	0	8
217	Isoform C2 of Heterogeneous nuclear ribonucleoproteins C1/C2	IPI00477313	3	11	0	8
218	Isoform Complexed of Arginyl-tRNA synthetase, cytoplasmic	IPI00004860	6	37	2	11
219	Isoform Cytoplasmic of Lysyl-tRNA synthetase	IPI00014238	4	34	2	9
220	Isoform F of Constitutive coactivator of PPAR-gamma-like protein 1	IPI00384265	1	7	0	5
221	Isoform GTBP-N of DNA mismatch repair protein Msh6	IPI00384456	21	72	1	26
222	Isoform Long of Delta-1-pyrroline-5-carboxylate synthetase	IPI00008982	0	20	1	12
223	Isoform Long of Eukaryotic translation initiation factor 4H	IPI00014263	7	66	5	23
224	Isoform Long of Splicing factor, proline- and glutamine-rich	IPI00010740	22	80	9	30
225	Isoform Long of Trifunctional purine biosynthetic protein adenosine-3	IPI00025273	25	95	1	18
226	Isoform M2 of Pyruvate kinase isozymes M1/M2	IPI00479186	166	549	59	194

(continued)

Table A.1 (continued)

No.	Protein description	Protein ID	Total spectral counts			
			Expt #1		Expt #2	
			Control	Probe	Control	Probe
227	Isoform Mitochondrial of Fumarate hydratase, mitochondrial	IPI00296053	17	70	5	24
228	Isoform Mitochondrial of Peroxiredoxin-5, mitochondrial	IPI00024915	12	50	2	26
229	Isoform XLas-1 of Guanine nucleotide-binding protein G(s) subunit alpha isoforms XLas	IPI00095891	1	49	1	35
230	Isoleucyl-tRNA synthetase, cytoplasmic	IPI00644127	23	63	1	30
231	lactate dehydrogenase A isoform 3	IPI00947127	246	828	54	144
232	Lactoylglutathione lyase	IPI00220766	4	57	2	27
233	Lamina-associated polypeptide 2, isoform alpha	IPI00216230	0	35	0	23
234	Leucine-rich PPR motif-containing protein, mitochondrial	IPI00783271	94	195	14	98
235	Leucine-rich repeat-containing protein 59	IPI00396321	2	12	2	7
236	Lupus La protein	IPI00009032	2	16	0	5
237	MACRO domain-containing protein 1	IPI00155601	3	24	1	14
238	Malate dehydrogenase	IPI00916111	9	25	3	7
239	Methionyl-tRNA synthetase, cytoplasmic	IPI00008240	29	64	1	5
240	Microtubule-associated protein 1S	IPI00296485	0	9	0	7
241	Microtubule-associated protein RP/EB family member 1	IPI00017596	12	27	3	8
242	Mitochondrial import inner membrane translocase subunit TIM44	IPI00306516	0	16	0	5
243	Moesin	IPI00219365	8	98	1	36
244	Monocarboxylate transporter 1	IPI00024650	9	26	3	8
245	Myosin-Ie	IPI00329672	9	24	1	11
246	N-acetyltransferase 10	IPI00300127	5	16	0	7
247	Neuroblast differentiation-associated protein AHNAK	IPI00021812	4	126	0	70
248	Non-POU domain-containing octamer-binding protein	IPI00304596	11	84	4	16
249	Nuclear migration protein nudC	IPI00550746	0	14	0	8
250	Nucleolin	IPI00604620	14	43	3	37
251	Nucleosome assembly protein 1-like 1	IPI00023860	8	41	0	10
252	Peptidyl-prolyl cis-trans isomerase A	IPI00419585	45	97	16	62
253	Peptidyl-prolyl cis-trans isomerase B	IPI00646304	11	48	2	22
254	Peptidyl-prolyl cis-trans isomerase, mitochondrial	IPI00026519	2	14	0	7

(continued)

Table A.1 (continued)

No.	Protein description	Protein ID	Total spectral counts			
			Expt #1		Expt #2	
			Control	Probe	Control	Probe
255	Peroxiredoxin-1	IPI00000874	72	271	26	81
256	Peroxiredoxin-2	IPI00027350	24	111	8	26
257	Peroxiredoxin-4	IPI00011937	16	40	1	20
258	Peroxiredoxin-6	IPI00220301	23	59	1	37
259	Phosphatidylethanolamine-binding protein 1	IPI00219446	21	71	5	38
260	Phosphoglycerate mutase 1	IPI00549725	41	270	10	68
261	Phospholipase D3	IPI00328243	0	21	0	22
262	Phosphoribosylformylglycinamidine synthase	IPI00929344	24	107	0	21
263	Plastin-3	IPI00216694	18	91	6	17
264	Poly [ADP-ribose] polymerase 1	IPI00449049	39	79	4	30
265	Poly A binding protein, cytoplasmic 4 isoform 1	IPI00642904	2	16	0	11
266	Poly(rC)-binding protein 1	IPI00016610	23	68	12	89
267	Polypyrimidine tract-binding protein 1 isoform a	IPI00183626	31	106	3	74
268	Probable ATP-dependent RNA helicase DDX5	IPI00017617	13	72	0	17
269	Profilin-1	IPI00216691	31	81	5	31
270	Proliferating cell nuclear antigen	IPI00021700	21	42	2	5
271	Prolyl endopeptidase	IPI00008164	0	31	0	29
272	Prostaglandin E synthase 3 (Cytosolic), isoform CRA_c	IPI00789101	5	14	0	7
273	Proteasomal ubiquitin receptor ADRM1	IPI00033030	0	8	0	5
274	Proteasome subunit alpha type-2	IPI00219622	12	42	1	17
275	Proteasome subunit alpha type-5	IPI00291922	9	39	2	10
276	Proteasome subunit alpha type-6	IPI00029623	6	56	0	21
277	Proteasome subunit beta type-2	IPI00028006	21	57	1	9
278	Proteasome subunit beta type-3	IPI00028004	5	23	0	6
279	Proteasome subunit beta type-4	IPI00555956	18	37	0	15
280	Proteasome subunit beta type-5	IPI00479306	17	88	1	18
281	Proteasome subunit beta type-6	IPI00000811	17	73	0	7
282	Proteasome subunit beta type-7	IPI00003217	12	33	1	10
283	Protein disulfide-isomerase A3	IPI00025252	11	61	6	34
284	Protein disulfide-isomerase	IPI00010796	21	56	6	34
285	Protein FADD	IPI00011919	0	14	0	6
286	Protein NDRG1	IPI00022078	5	37	11	37
287	Protein RCC2	IPI00465044	0	42	1	13

(continued)

Table A.1 (continued)

No.	Protein description	Protein ID	Total spectral counts			
			Expt #1		Expt #2	
			Control	Probe	Control	Probe
288	Protein S100-A10	IPI00183695	9	33	0	7
289	Protein S100-A11	IPI00013895	23	179	14	41
290	Protein transport protein Sec24C	IPI00024661	5	39	3	25
291	Pumilio homolog 1 (Drosophila), isoform CRA_c	IPI00032355	1	5	0	8
292	Putative pre-mRNA-splicing factor ATP-dependent RNA helicase DHX15	IPI00396435	1	49	0	7
293	Cell division cycle 2 isoform 3	IPI00026689	2	23	3	8
294	Host cell factor	IPI00641743	0	14	0	23
295	Hypoxanthine-guanine phosphoribosyltransferase	IPI00873466	12	29	2	13
296	Phosphatidylinositol-binding clathrin assembly protein	IPI00871890	0	42	0	23
297	Ran-specific GTPase-activating protein	IPI00878611	0	11	0	6
298	60S ribosomal protein L30	IPI00872940	6	13	0	5
299	Ras GTPase-activating protein-binding protein 1	IPI00012442	3	31	0	13
300	Ras-related protein Rab-1B	IPI00008964	2	17	0	22
301	Ras-related protein Rab-7a	IPI00016342	5	25	0	17
302	Ribonucleases P/MRP protein subunit POP1	IPI00293331	0	7	1	5
303	Ribonucleoside-diphosphate reductase large subunit	IPI00013871	13	103	1	19
304	Ribosomal protein L14 variant	IPI00555744	0	10	0	7
305	Ribosyldihydronicotinamide dehydrogenase [quinone]	IPI00219129	2	71	0	29
306	SEC24B protein	IPI00030851	1	16	0	14
307	Serine hydroxymethyltransferase, mitochondrial	IPI00002520	7	82	3	43
308	Serine/threonine-protein kinase N2	IPI00002804	2	8	1	5
309	Serine/threonine-protein phosphatase 2A 65 kDa regulatory subunit A alpha isoform	IPI00554737	14	68	3	15
310	Serine/threonine-protein phosphatase PP1-beta catalytic subunit	IPI00218236	7	59	0	6
311	SH2 domain-containing protein 4A	IPI00100731	0	7	0	9
312	Sorbitol dehydrogenase	IPI00216057	8	30	2	5
313	Splicing factor 3 subunit 1	IPI00017451	2	27	0	5
314	Splicing factor 3B subunit 1	IPI00026089	15	68	0	27
315	Splicing factor 3B subunit 2	IPI00221106	9	35	0	13
316	Staphylococcal nuclease domain-containing protein 1	IPI00140420	49	101	4	11

(continued)

Table A.1 (continued)

No.	Protein description	Protein ID	Total spectral counts			
			Expt #1		Expt #2	
			Control	Probe	Control	Probe
317	Structural maintenance of chromosomes protein 1A	IPI00291939	6	15	0	14
318	Structural maintenance of chromosomes protein 3	IPI00219420	1	13	0	10
319	Superkiller viralicidic activity 2-like 2	IPI00647217	2	22	2	8
320	Superoxide dismutase [Mn], mitochondrial	IPI00022314	3	22	0	14
321	SWI/SNF complex subunit SMARCC1	IPI00234252	4	13	0	7
322	Talin-1	IPI00298994	24	89	2	32
323	T-complex protein 1 subunit alpha	IPI00290566	20	100	5	18
324	T-complex protein 1 subunit beta	IPI00297779	5	64	1	18
325	T-complex protein 1 subunit delta	IPI00302927	6	76	0	22
326	T-complex protein 1 subunit epsilon	IPI00010720	20	124	3	16
327	T-complex protein 1 subunit eta	IPI00018465	15	86	2	16
328	T-complex protein 1 subunit gamma	IPI00553185	8	97	2	16
329	T-complex protein 1 subunit theta	IPI00784090	23	96	10	30
330	T-complex protein 1 subunit zeta	IPI00027626	6	64	2	15
331	TAR DNA binding protein 43	IPI00025815	3	29	0	14
332	Thimet oligopeptidase	IPI00549189	2	29	0	5
333	Thioredoxin	IPI00216298	17	142	1	24
334	Thioredoxin-dependent peroxide reductase, mitochondrial	IPI00024919	6	24	2	7
335	Threonyl-tRNA synthetase, cytoplasmic	IPI00329633	28	109	5	21
336	Transcription factor A, mitochondrial	IPI00020928	0	11	0	5
337	Transcriptional repressor p66-beta	IPI00103554	0	10	0	5
338	Transforming protein RhoA	IPI00478231	9	20	1	5
339	Transitional endoplasmic reticulum ATPase	IPI00022774	23	102	7	19
340	Translin	IPI00018768	4	18	0	7
341	Transmembrane emp24 domain-containing protein 10	IPI00028055	6	17	1	5
342	Transmembrane protein 33	IPI00299084	6	24	2	8
343	Trifunctional enzyme subunit alpha, mitochondrial	IPI00031522	3	40	0	17
344	Triosephosphate isomerase 1 isoform 2	IPI00465028	41	235	29	94
345	Tu translation elongation factor, mitochondrial precursor	IPI00027107	21	48	2	21
346	Tubulin-specific chaperone A	IPI00217236	5	23	1	8
347	Tumor protein D52-like 2 isoform a	IPI00399265	0	35	2	16
348	Tumor protein, translationally-controlled 1	IPI00009943	15	38	0	19

(continued)

Table A.1 (continued)

No.	Protein description	Protein ID	Total spectral counts			
			Expt #1		Expt #2	
			Control	Probe	Control	Probe
349	Ubiquilin-2	IPI00409659	0	7	0	13
350	Ubiquitin carboxyl-terminal hydrolase 7	IPI00003965	7	21	0	5
351	Ubiquitin-associated protein 2	IPI00171127	2	25	0	17
352	Ubiquitin-conjugating enzyme E2 C	IPI00013002	4	9	2	5
353	UMP-CMP kinase 1 isoform a	IPI00219953	1	17	2	6
354	UPF0568 protein C14orf166	IPI00006980	2	23	0	9
355	Vacuolar protein sorting-associated protein 35	IPI00018931	4	66	2	9
356	Valyl-tRNA synthetase	IPI00893918	41	112	0	33
357	Very long-chain acyl-CoA synthetase	IPI00024787	7	36	2	14
358	Vesicle-trafficking protein SEC22b	IPI00006865	9	40	5	21
359	Von Hippel-Lindau binding protein 1	IPI00334159	0	20	0	6
360	Zinc transporter 1	IPI00002483	0	25	0	16
361	Zyxin	IPI00926625	0	10	0	12

Each protein listed has been identified in the both independent experiments with support from at least 2 unique peptides, total spectral counts greater or equal to 5, and the ratio of MalAM-yne/control greater than 2

H4K5 $_{pr}$GK$_{glu}$GGK$_{ac}$GLGK$_{ac}$GGAK$_{ac}$R
Charge: 2 m/z: 783.9336

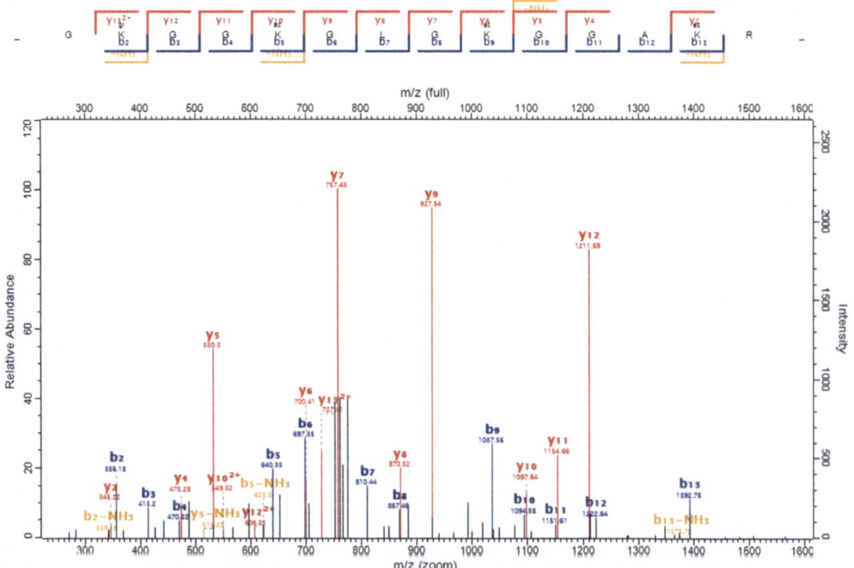

b ion					y ion		y²⁺ ion	
Δ dalton	mass		seq		mass	△ dalton	mass	△ dalton
	114.055	1	G	13				
+0.14544	356.182	2	K	12	1453.81		727.41	+0.0746
+0.13085	413.203	3	G	11	1211.69	-0.0236	606.346	+0.17327
+0.12592	470.225	4	G	10	1154.66	+0.01321	1154.66	
+0.0608	640.33	5	K	9	1097.64	+0.068	549.325	+0.0488
+0.06943	697.352	6	G	8	927.537	+0.07935	927.537	
+0.08784	810.436	7	L	7	870.516	+0.06993	870.516	
+0.08017	867.457	8	G	6	757.432	+0.08063	757.432	
+0.017	1037.56	9	K	5	700.41	+0.05284	700.41	
+0.1275	1094.58	10	G	4	530.305	+0.00456	530.305	
+0.10591	1151.61	11	G	3	473.283	+0.08102	473.283	
+0.09175	1222.64	12	A	2	416.262		416.262	
+0.07973	1392.75	13	K	1	345.224	+0.1012	345.224	
		14	R	0	175.119		175.119	

H4K12 $_{pr}$GLGK$_{glu}$GGAK$_{ac}$R

Charge: 2 m/z: 528.2958

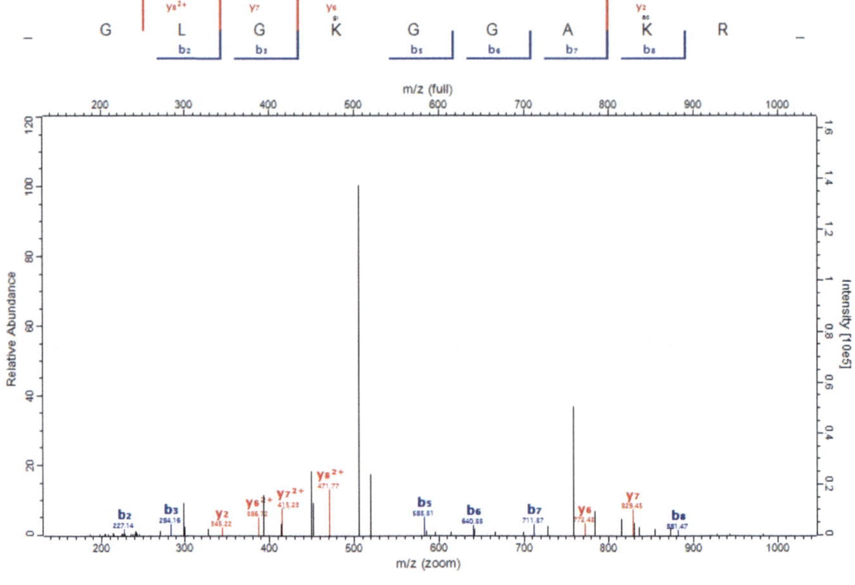

b ion Δ dalton	b ion mass		seq		y ion mass	y ion △ dalton	y²⁺ ion mass	y²⁺ ion △ dalton
	114.055	1	G	8				
-0.0234	227.139	2	L	7	942.537		471.772	+0.18272
+0.12455	284.16	3	G	6	829.453	+0.0343	415.23	+0.12557
	526.287	4	K	5	772.431	+0.02543	386.719	+0.29377
-0.1174	583.309	5	G	4	530.305		530.305	
+0.02961	640.33	6	G	3	473.283		473.283	
-0.0568	711.367	7	A	2	416.262		416.262	
+0.055	881.473	8	K	1	345.224	+0.11344	345.224	
		9	R	0	175.119		175.119	

H4K31 $_{pr}$DNIQGITK$_{glu}$PAIR

Charge: 2 m/z: 748.4094

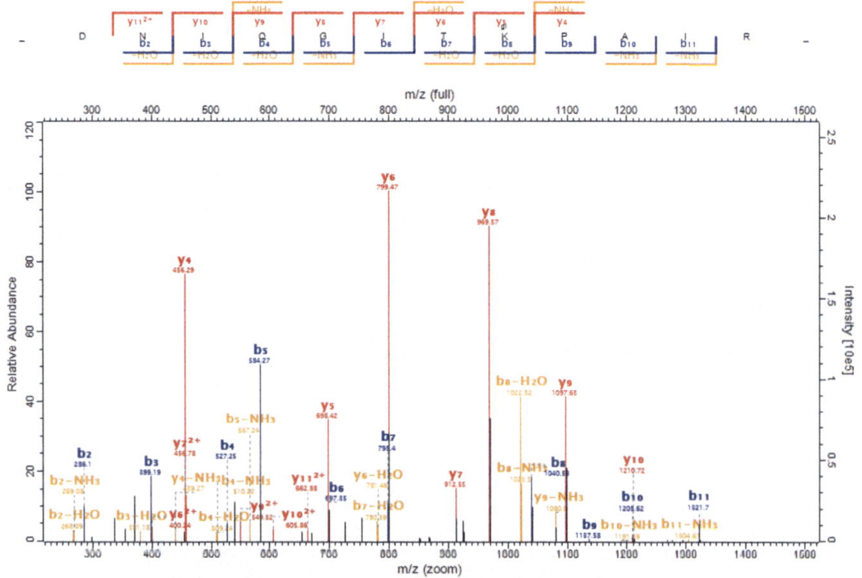

b ion					y ion		y²⁺ ion	
Δ dalton	mass		seq		mass	△ dalton	mass	△ dalton
	172.06	1	D	11				
+0.03534	286.103	2	N	10	1324.76		662.883	+0.07521
-0.0111	399.187	3	I	9	1210.72	-0.012	605.861	+0.13189
-0.2278	527.246	4	Q	8	1097.63	-0.0553	549.319	+0.04141
-0.0638	584.267	5	G	7	969.573	-0.022	969.573	
+0.01047	697.352	6	I	6	912.551	+0.00773	456.779	+0.48567
+0.04104	798.399	7	T	5	799.467	+0.0426	400.237	+0.0114
-0.0362	1040.53	8	K	4	698.42	+0.0207	698.42	
-0.0237	1137.58	9	P	3	456.293	-0.0073	456.293	
-0.0604	1208.62	10	A	2	359.24		359.24	
-0.073	1321.7	11	I	1	288.203		288.203	
		12	R	0	175.119		175.119	

H4K59 $_{pr}$GVLK$_{glu}$VFLENVIR
Charge: 2 m/z: 778.9560

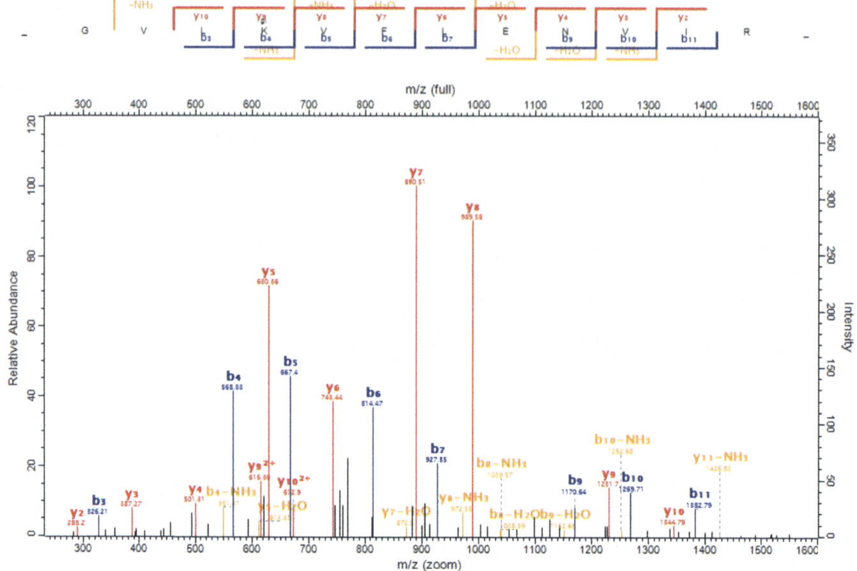

b ion					y ion		y²⁺ ion	
Δ dalton	mass		seq		mass	△ dalton	mass	△ dalton
	114.055	1	G	11				
	213.123	2	V	10	1443.86		1443.86	
+0.18453	326.207	3	L	9	1344.79	+0.13099	672.898	+0.05142
-0.00242	568.334	4	K	8	1231.7	-0.00125	616.356	+0.06721
+0.01113	667.403	5	V	7	989.578	+0.04264	989.578	
+0.03769	814.471	6	F	6	890.509	+0.00449	890.509	
+0.05976	927.555	7	L	5	743.441	+0.07614	743.441	
	1056.6	8	E	4	630.357	+0.02226	630.357	
+0.10754	1170.64	9	N	3	501.314	+0.02671	501.314	
+0.05769	1269.71	10	V	2	387.271	+0.09649	387.271	
-0.10231	1382.79	11	I	1	288.203	-0.09962	288.203	
		12	R	0	175.119		175.119	

H4K77 $_{pr}$DAVTYTEHAK$_{glu}$R

Charge: 2 m/z: 730.8544

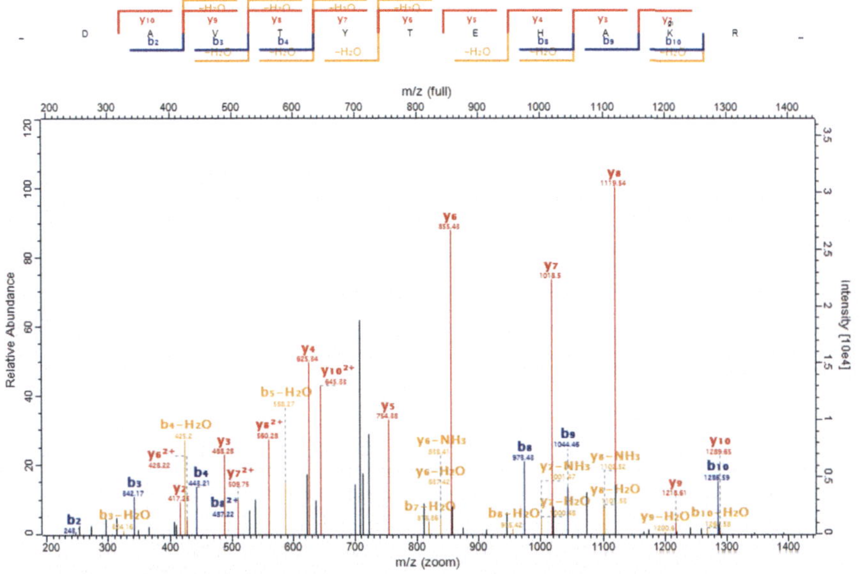

b²⁺ ion		b ion					y ion		y²⁺ ion	
Δ dalton	mass	Δ dalton	mass		seq		mass	△ dalton	mass	△ dalton
	172.06		172.06	1	D	10				
	243.1	+0.1124	243.1	2	A	9	1289.6	+0.0471	645.33	+0.0102
	342.17	+0.0739	342.17	3	V	8	1218.6	-0.041	1218.6	
	443.21	+0.0182	443.21	4	T	7	1119.5	-0.048	560.28	-0.039
	606.28		606.28	5	Y	6	1018.5	-0.025	509.75	-0.117
	707.32		707.32	6	T	5	855.43	-0.004	428.22	+0.111
	836.37		836.37	7	E	4	754.38	+0.0022	754.38	
+0.0437	487.22	-0.004	973.43	8	H	3	625.34	+4E-05	625.34	
	1044.5	-0.032	1044.5	9	A	2	488.28	-0.004	488.28	
	1286.6	-0.111	1286.6	10	K	1	417.25	+0.0268	417.25	
				11	R	0	175.12		175.12	

H4K79 $_{pr}K_{glu}$TVTAMDVVYALK$_{glu}$R

Charge: 2 m/z: 939.9977

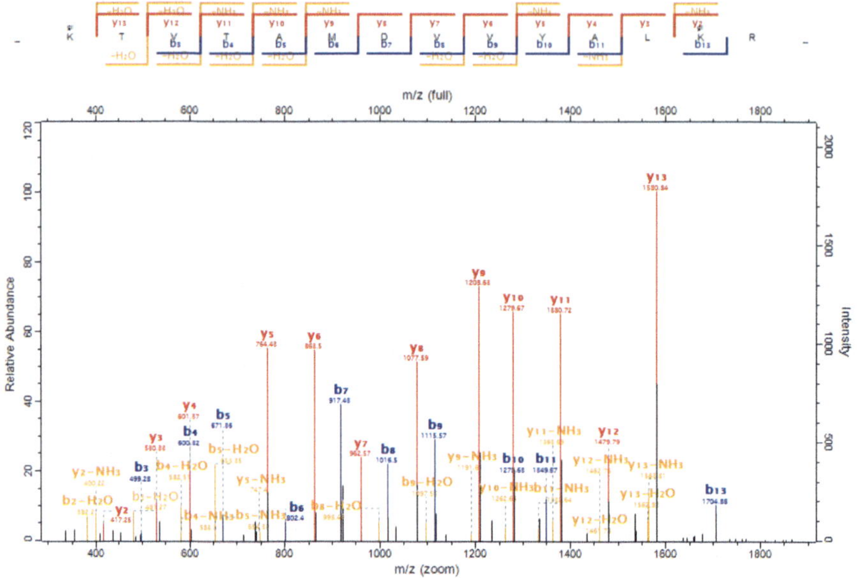

b ion					y ion	
Δ dalton	mass		seq		mass	△ dalton
	299.1601483	1	K	13		
	400.2078268	2	T	12	1580.835256	-0.0260516
+0.0658614	499.2762407	3	V	11	1479.787577	+0.0029501
+0.0670721	600.3239192	4	T	10	1380.719163	-0.0524152
-0.0116677	671.3610329	5	A	9	1279.671485	-0.0456303
+0.1178916	802.4015176	6	M	8	1208.634371	-0.0697958
+0.185431	917.4284606	7	D	7	1077.593886	+0.0102395
-0.0353266	1016.496875	8	V	6	962.5669434	+0.0443847
-0.0311087	1115.565288	9	V	5	863.4985295	+0.0557918
-0.0863806	1278.628617	10	Y	4	764.4301156	+0.041381
+0.0260417	1349.665731	11	A	3	601.366787	-0.0280419
	1462.749795	12	L	2	530.3296732	-0.0196146
-0.0108512	1704.876452	13	K	1	417.2456093	+0.1851769
		14	R	0	175.1189522	

H4K91 $_{pr}$K$_{pr}$TVTAMDVVYALK$_{glu}$R

Charge: 2 m/z: 910.99496

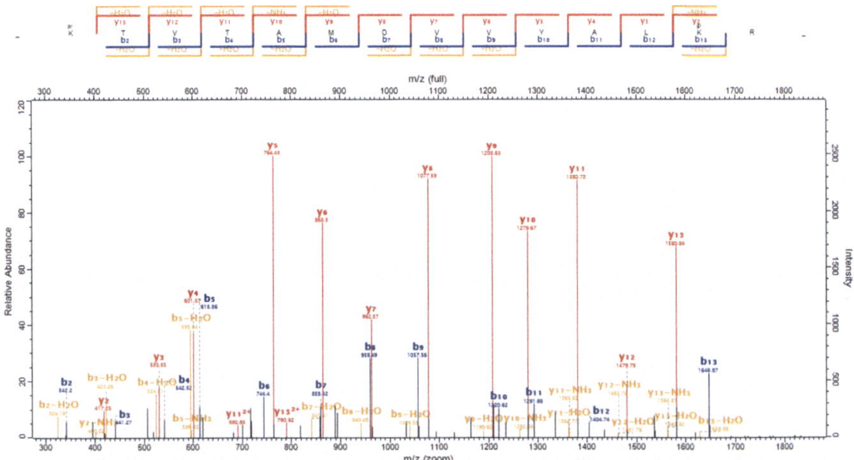

b ion					y ion		y^{2+} ion	
Δ dalton	mass		seq		mass	△ dalton	mass	△ dalton
	241.155	1	K	13				
+0.06279	342.202	2	T	12	1580.84	-0.0743	790.921	-0.0163
+0.01769	441.271	3	V	11	1479.79	-0.0476	1479.79	
-0.0663	542.318	4	T	10	1380.72	-0.0939	690.863	+0.073
-0.0804	613.356	5	A	9	1279.67	-0.0656	1279.67	
-0.05	744.396	6	M	8	1208.63	-0.0746	1208.63	
-0.0407	859.423	7	D	7	1077.59	-0.002	1077.59	
-0.0379	958.491	8	V	6	962.567	+0.0168	962.567	
-0.0339	1057.56	9	V	5	863.499	+0.05237	863.499	
-0.0865	1220.62	10	Y	4	764.43	+0.0152	764.43	
-0.0587	1291.66	11	A	3	601.367	-0.0471	601.367	
-0.1359	1404.74	12	L	2	530.33	+0.02177	530.33	
-0.1019	1646.87	13	K	1	417.246	+0.01355	417.246	
		14	R	0	175.119		175.119	

H3K14 $_{pr}$K$_{pr}$STGGK$_{glu}$APR
Charge: 2 m/z: 564.3064

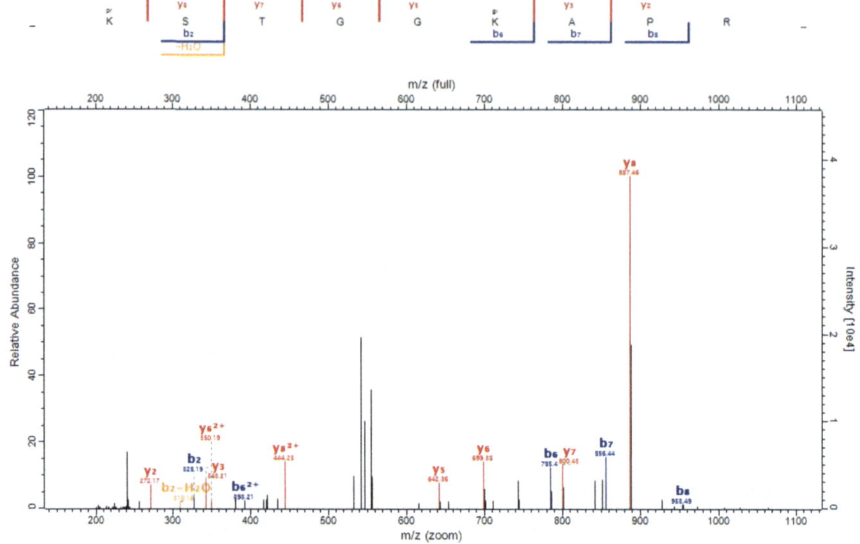

b²⁺ ion		b ion					y ion		y²⁺ ion	
△ dalton	mass	△ dalton	mass		seq		mass	△ dalton	mass	△ dalton
	241.15		241.15	1	K	8				
	328.19	-0.047	328.19	2	S	7	887.46	+0.0335	444.23	+0.122
	429.23		429.23	3	T	6	800.43	+0.0541	800.43	
	486.26		486.26	4	G	5	699.38	+0.0274	350.19	+0.2611
	543.28		543.28	5	G	4	642.36	+0.0077	642.36	
+0.1077	393.21	-0.015	785.4	6	K	3	585.34		585.34	
	856.44	+0.038	856.44	7	A	2	343.21	+0.0843	343.21	
	953.49	+0.1431	953.49	8	P	1	272.17	+0.0753	272.17	
				9	R	0	175.12		175.12	

H3K18 $_{pr}$K$_{glu}$QLATK$_{ac}$AAR

Charge: 2 m/z: 599.8431

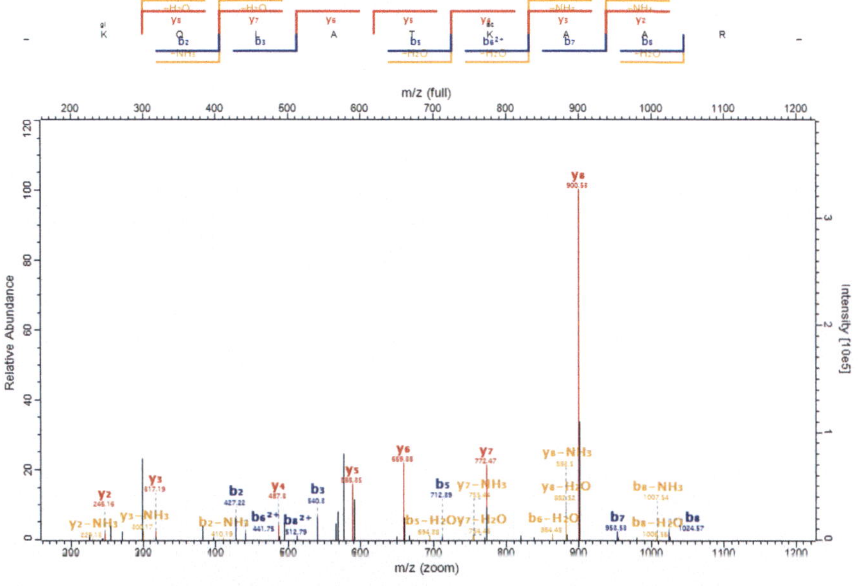

b²⁺ ion		b ion					y ion	
Δ dalton	mass	Δ dalton	mass		seq		mass	△ dalton
	299.16		299.16	1	K	8		
	427.219	+0.09152	427.219	2	Q	7	900.526	+0.04759
	540.303	-0.05578	540.303	3	L	6	772.468	+0.04373
	611.34		611.34	4	A	5	659.383	+0.08769
	712.388	+0.15465	712.388	5	T	4	588.346	+0.04125
+0.15301	441.75		882.493	6	K	3	487.299	+0.11822
	953.53	+0.07329	953.53	7	A	2	317.193	+0.18109
-0.43733	512.787	+0.1425	1024.57	8	A	1	246.156	+0.12343
				9	R	0	175.119	

H3K23 $_{pr}$K$_{pr}$QLATK$_{glu}$AAR

Charge: 2 m/z: 606.8510

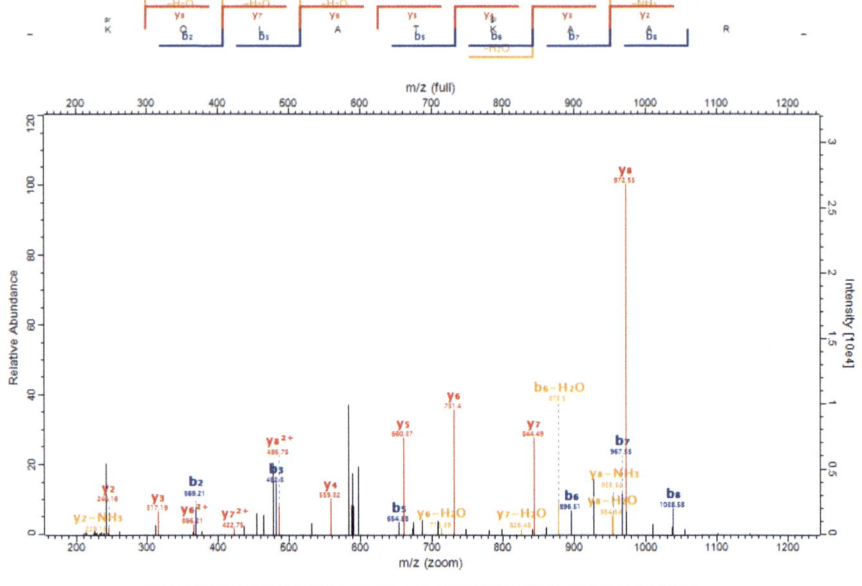

b ion					y ion		y^{2+} ion	
Δ dalton	mass		seq		mass	△ dalton	mass	△ dalton
	241.155	1	K	8				
+0.0218	369.213	2	Q	7	972.547	-0.0156	486.777	-0.0697
-0.0128	482.297	3	L	6	844.489	+0.00905	422.748	+0.46088
	553.334	4	A	5	731.405	+0.01505	366.206	+0.13737
+0.06742	654.382	5	T	4	660.368	+0.06248	660.368	
+0.08621	896.509	6	K	3	559.32	-0.0471	559.32	
+0.10348	967.546	7	A	2	317.193	+0.02407	317.193	
-0.067	1038.58	8	A	1	246.156	+0.0687	246.156	
		9	R	0	175.119		175.119	

H3K27 $_{pr}$K$_{glu}$SAPATGGVK$_{pr}$
Charge: 2 m/z: 571.3086

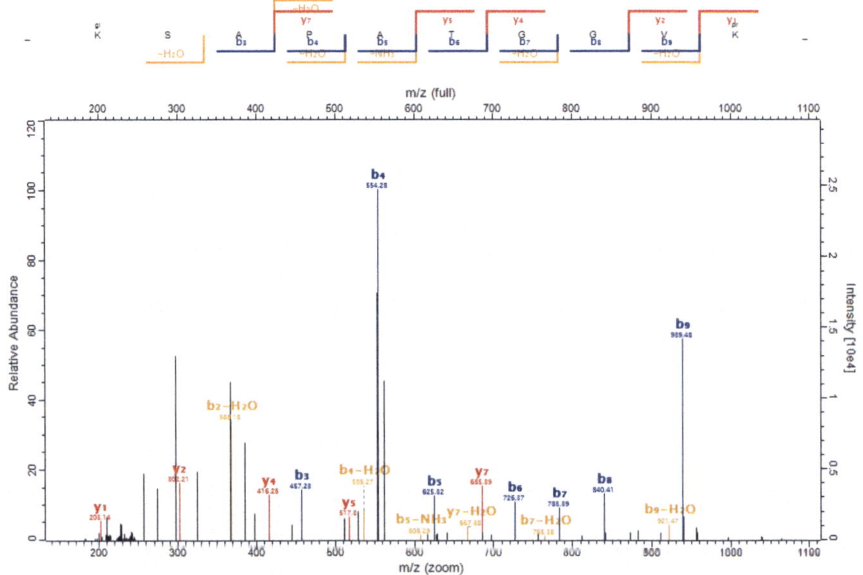

b ion					y ion	
Δ dalton	mass		seq		mass	△ dalton
	299.16014829	1	K	9		
	386.1921767	2	S	8	843.4570586	
+0.0573916	457.22929049	3	A	7	756.42503019	
-0.0275988	554.28205434	4	P	6	685.3879164	-0.0212172
+0.0355072	625.31916813	5	A	5	588.33515255	
+0.1086173	726.36684661	6	T	4	517.29803876	+0.0046346
+0.0042068	783.38831033	7	G	3	416.25036028	+0.0918333
+0.0216835	840.40977405	8	G	2	359.22889656	
-0.0038838	939.47818797	9	V	1	302.20743284	+0.0818738
		10	K	0	203.13901892	+0.0492898

H3K56 $_{pr}$RYQK$_{glu}$STELLIR

Charge: 2 m/z: 788.9383

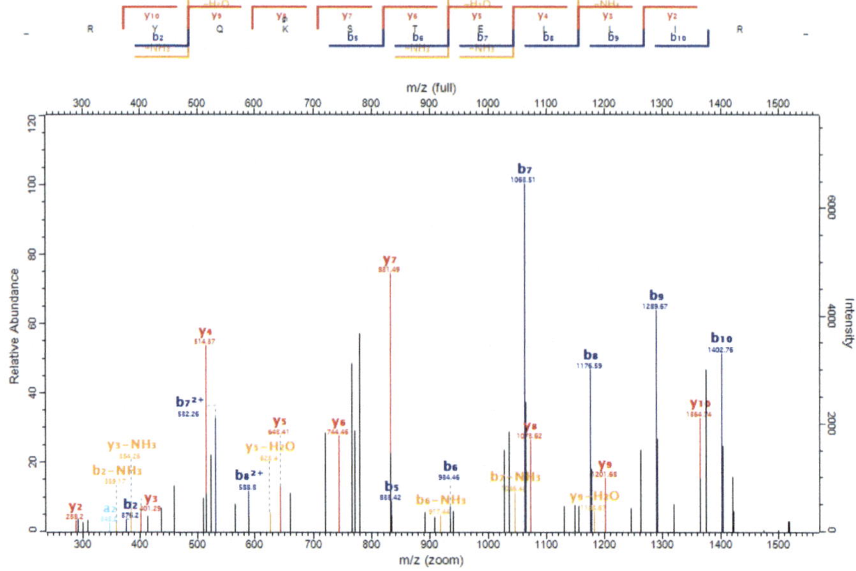

b²⁺ ion		b ion					y ion	
Δ dalton	mass	Δ dalton	mass		seq		mass	△ dalton
	213.13		213.13	1	R	10		
	376.2	+0.1383	376.2	2	Y	9	1364.7	-0.025
	504.26		504.26	3	Q	8	1201.7	-0.048
	746.38		746.38	4	K	7	1073.6	+0.0094
	833.42	+0.0376	833.42	5	S	6	831.49	+0.0421
	934.46	+0.0533	934.46	6	T	5	744.46	+0.0195
+0.074	532.26	+0.0286	1063.5	7	E	4	643.41	-0.01
-0.054	588.8	+0.001	1176.6	8	L	3	514.37	-0.039
	1289.7	-0.038	1289.7	9	L	2	401.29	+0.0694
	1402.8	-0.017	1402.8	10	I	1	288.2	+0.0421
				11	R	0	175.12	

H3K79 $_{pr}$EIAQDFK$_{glu}$TDLR

Charge: 3 m/z: 753.3778

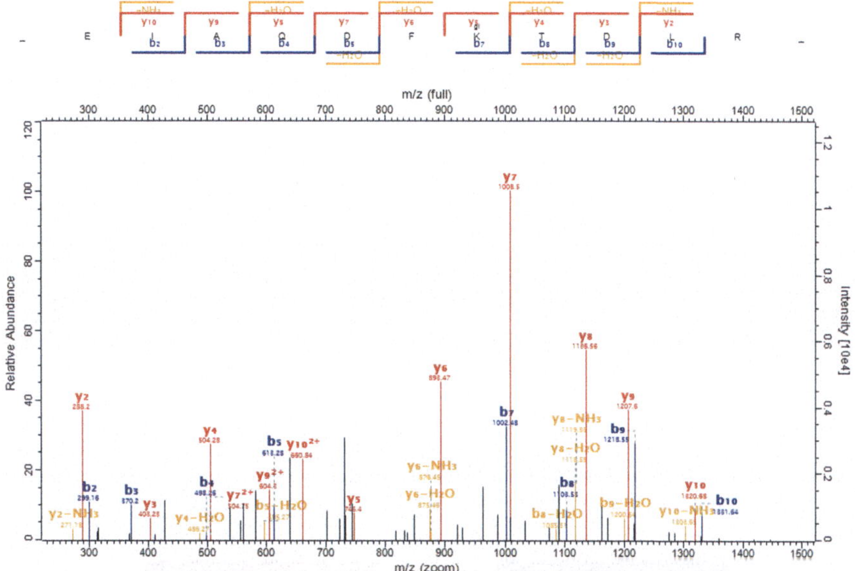

b ion					y ion		y²⁺ ion	
Δ dalton	mass		seq		mass	△ dalton	mass	△ dalton
	186.076	1	E	10				
+0.1088	299.16	2	I	9	1320.68	-0.04684	660.843	-0.01412
+0.0915	370.197	3	A	8	1207.6	-0.0691	604.301	+0.05074
+0.09328	498.256	4	Q	7	1136.56	-0.06861	1136.56	
-0.04151	613.283	5	D	6	1008.5	-0.01479	504.753	+0.40327
	760.351	6	F	5	893.473	-0.02953	893.473	
-0.00575	1002.48	7	K	4	746.404	+0.02741	746.404	
-0.08461	1103.53	8	T	3	504.278	-0.0371	504.278	
-0.10594	1218.55	9	D	2	403.23	+0.0754	403.23	
-0.08734	1331.64	10	L	1	288.203	+0.01183	288.203	
		11	R	0	175.119		175.119	

H3K115 FQSAAIGALQEASEAYLVGLFEDTNLCAIHAK$_{glu}$R

Charge: 4 m/z: 927.7148

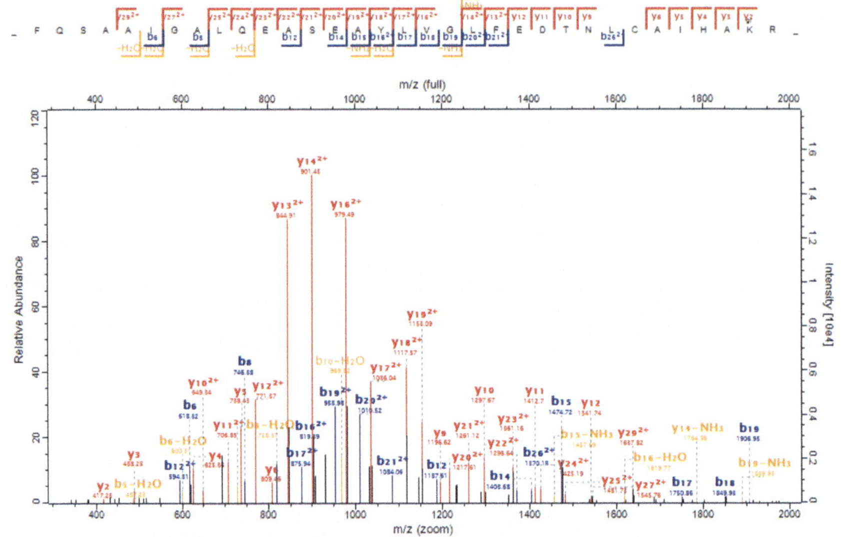

b²⁺ ion		b ion					y ion		y²⁺ ion	
Δ dalton	mass	Δ dalton	mass		seq		mass	△ dalton	mass	△ dalton
	148.08		148.08	1	F	32				
	276.13		276.13	2	Q	31	3560.8		3560.8	
	363.17		363.17	3	S	30	3432.7		3432.7	
	434.2		434.2	4	A	29	3345.7		3345.7	
	505.24		505.24	5	A	28	3274.6		1637.8	-0.005
	618.32	+0.0792	618.32	6	I	27	3203.6		3203.6	
	675.35		675.35	7	G	26	3090.5		1545.8	+0.4943
	746.38	+0.069	746.38	8	A	25	3033.5		3033.5	
	859.47		859.47	9	L	24	2962.5		1481.7	+0.4261
	987.53		987.53	10	Q	23	2849.4		1425.2	+0.3657
	1116.6		1116.6	11	E	22	2721.3		1361.2	+0.2926
+0.0949	594.31	-0.216	1187.6	12	A	21	2592.3		1296.6	+0.1966
	1274.6		1274.6	13	S	20	2521.2		1261.1	+0.1691
	1403.7	-0.252	1403.7	14	E	19	2434.2		1217.6	+0.1573
	1474.7	-0.036	1474.7	15	A	18	2305.2		1153.1	+0.2806
+0.2639	819.39		1637.8	16	Y	17	2234.1		1117.6	+0.24
+0.2636	875.94	-0.007	1750.9	17	L	16	2071.1		1036	+0.2955
	1849.9	+0.049	1849.9	18	V	15	1958		979.49	+0.293
+0.2703	953.98	+0.0902	1907	19	G	14	1858.9		1858.9	
+0.276	1010.5		2020	20	L	13	1801.9		901.45	+0.2991
-0.198	1084.1		2167.1	21	F	12	1688.8		844.91	+0.3198
	2296.1		2296.1	22	E	11	1541.7	+0.0603	771.37	+0.2371
	2411.2		2411.2	23	D	10	1412.7	+0.0438	706.85	+0.2326
	2512.2		2512.2	24	T	9	1297.7	+0.0166	649.34	+0.1064
	2626.3		2626.3	25	N	8	1196.6	+0.0067	1196.6	
+0.3509	1370.2		2739.4	26	L	7	1082.6		1082.6	
	2899.4		2899.4	27	C	6	969.49		969.49	
	2970.4		2970.4	28	A	5	809.46	+0.0636	809.46	
	3083.5		3083.5	29	I	4	738.43	-0.279	738.43	
	3220.6		3220.6	30	H	3	625.34	+0.1152	625.34	
	3291.6		3291.6	31	A	2	488.28	+0.0676	488.28	
	3533.7		3533.7	32	K	1	417.25	+0.1844	417.25	
				33	R	0	175.12		175.12	

H3K122 $_{pr}$VTIMPK$_{glu}$DIQLAR
Charge: 2 m/z: 777.9316

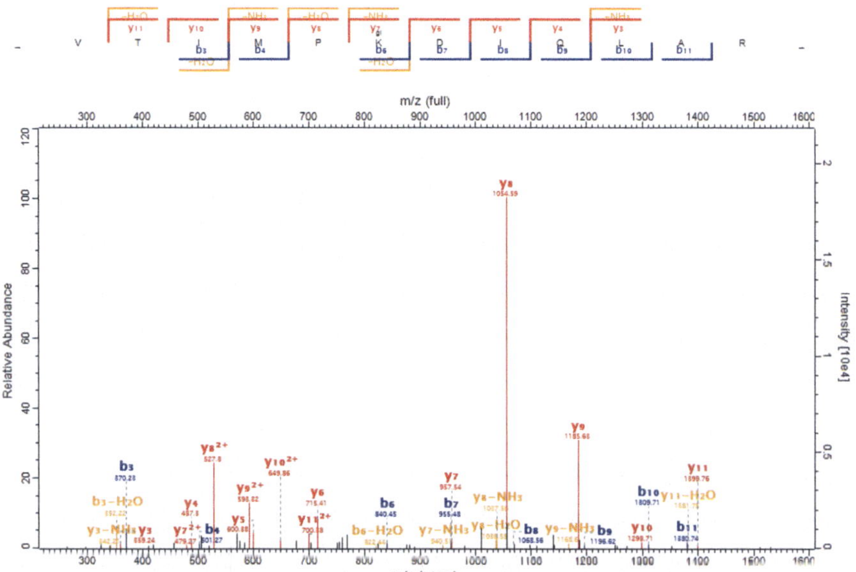

b ion					y ion		y^{2+} ion	
Δ dalton	mass		seq		mass	△ dalton	mass	△ dalton
	156.102	1	V	11				
	257.15	2	T	10	1399.76	-0.01734	700.384	-0.03221
+0.04846	370.234	3	I	9	1298.71	+0.07135	649.86	+0.02789
+0.12485	501.274	4	M	8	1185.63	+0.02077	593.318	+0.05649
	598.327	5	P	7	1054.59	+0.04514	527.798	+0.00428
+0.13678	840.454	6	K	6	957.536	+0.06513	479.272	+0.10125
+0.17441	955.48	7	D	5	715.41	+0.06038	715.41	
+0.03957	1068.56	8	I	4	600.383	+0.06803	600.383	
+0.10526	1196.62	9	Q	3	487.299	+0.00088	487.299	
+0.07234	1309.71	10	L	2	359.24	+0.22383	359.24	
+0.00556	1380.74	11	A	1	246.156		246.156	
		12	R	0	175.119		175.119	

H2BK16 $_{pr}K_{glu}AVTK_{ac}AQK_{pr}$
Charge: 2 m/z: 571.3268

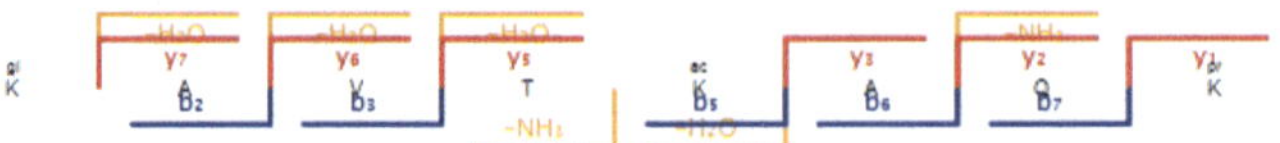

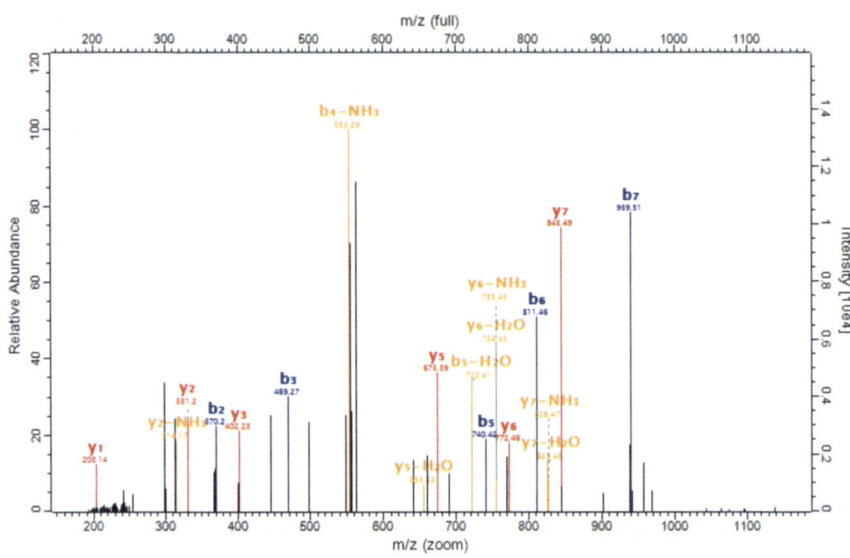

b ion					y ion	
Δ dalton	mass		seq		mass	△ dalton
	299.16014829	1	K	7		
+0.138889	370.19726208	2	A	6	843.4934441	+0.0562385
+0.0584817	469.265676	3	V	5	772.45633031	+0.0628347
	570.31335447	4	T	4	673.3879164	+0.0039903
+0.1000387	740.41888218	5	K	3	572.34023792	
+0.0155006	811.45599596	6	A	2	402.23471022	+0.0831914
+0.0378557	939.51457347	7	Q	1	331.19759643	+0.1884509
		8	K	0	203.13901892	-0.0268516

H2BK34 $_{pr}$SRK$_{glu}$ESYSVYVYK$_{pr}$

Charge: 2 m/z: 867.9329

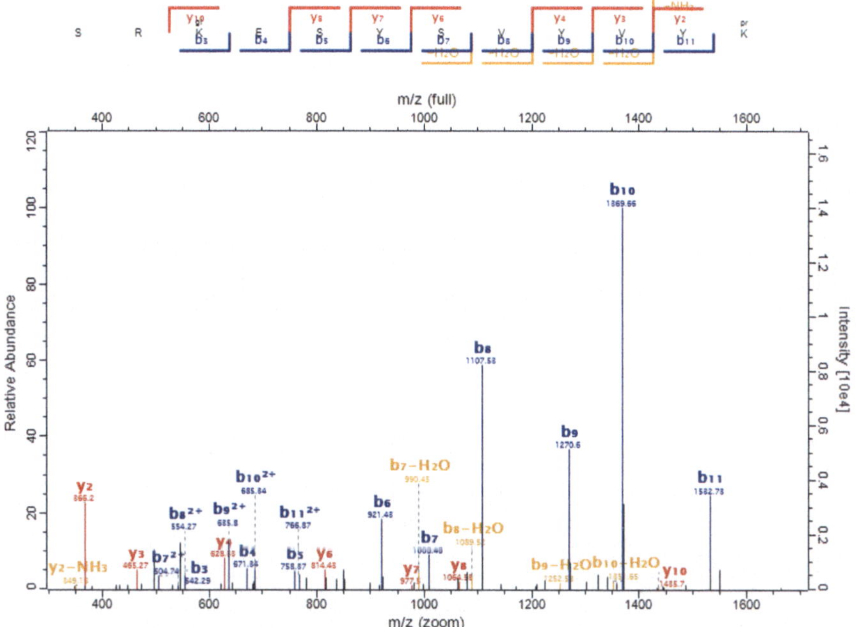

b²⁺ ion		b ion					y ion	
Δ dalton	mass	Δ dalton	mass		seq		mass	Δ dalton
	144.066		144.066	1	S	11		
	300.167		300.167	2	R	10	1591.8	
	542.293	-0.1392	542.293	3	K	9	1435.7	-0.1257
	671.336	+0.06793	671.336	4	E	8	1193.57	
	758.368	-0.0375	758.368	5	S	7	1064.53	-0.0202
	921.431	+0.01774	921.431	6	Y	6	977.498	+0.04645
-0.1223	504.735	+0.04687	1008.46	7	S	5	814.435	-0.0483
+0.00713	554.269	-0.0229	1107.53	8	V	4	727.403	
+0.0492	635.801	-0.0511	1270.6	9	Y	3	628.334	+0.01003
+0.08866	685.335	-0.0029	1369.66	10	V	2	465.271	-0.004
+0.06927	766.867	-0.0389	1532.73	11	Y	1	366.202	+0.06499
				12	K	0	203.139	

H2BK43 $_{pr}$ESYSVYVYK$_{glu}$VLK$_{pr}$

Charge: 2 m/z: 852.4426

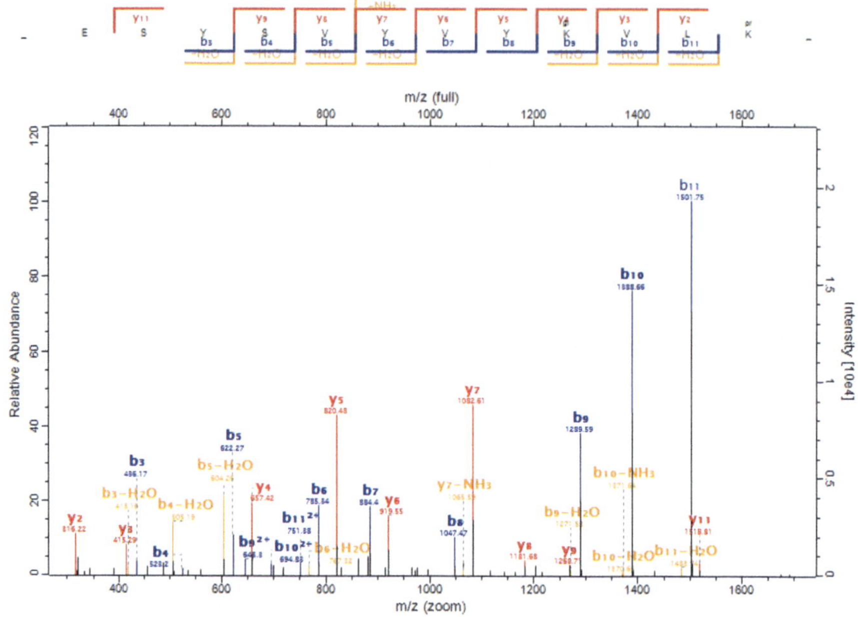

b²⁺ ion		b ion					y ion	
Δ dalton	mass	Δ dalton	mass		seq		mass	△ dalton
	186.076		186.076	1	E	11		
	273.108		273.108	2	S	10	1518.81	-0.00946
	436.171	+0.00757	436.171	3	Y	9	1431.78	
	523.203	+0.08022	523.203	4	S	8	1268.71	-0.05363
	622.272	+0.03653	622.272	5	V	7	1181.68	+0.12854
	785.335	+0.08227	785.335	6	Y	6	1082.61	+0.00946
	884.404	-0.02118	884.404	7	V	5	919.55	+0.10171
	1047.47	-0.03702	1047.47	8	Y	4	820.481	+0.08968
-0.07712	645.3	-0.03709	1289.59	9	K	3	657.418	+0.06183
+0.10248	694.835	-0.00834	1388.66	10	V	2	415.291	+0.11024
+0.24545	751.377	-0.00952	1501.75	11	L	1	316.223	+0.13654
				12	K	0	203.139	

H2BK46 $_{pr}$VLK$_{glu}$QVHPDTGISSK

Charge: 2 m/z: 839.9542

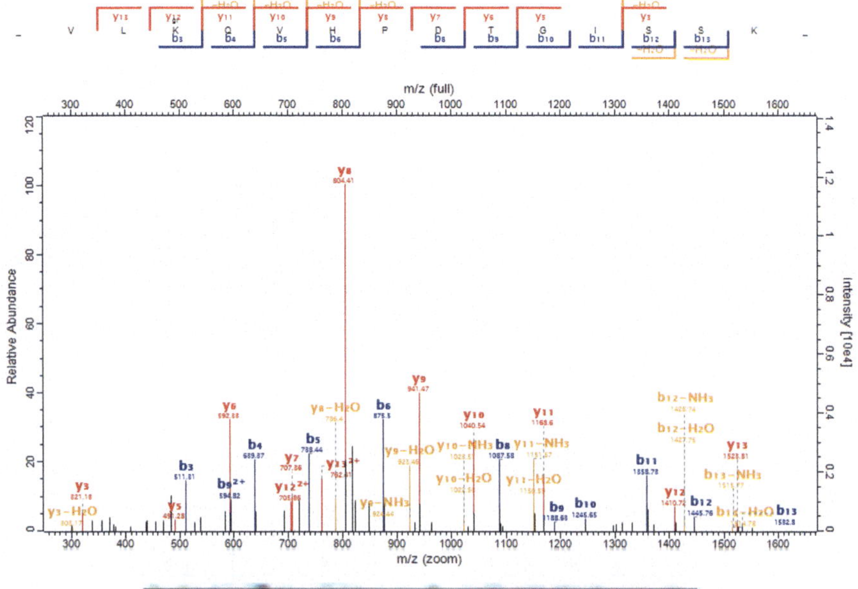

b²⁺ ion		b ion					y ion		y²⁺ ion	
Δ dalton	mass	Δ dalton	mass		seq		mass	△ dalton	mass	△ dalton
	156.1		156.1	1	V	13				
	269.19		269.19	2	L	12	1523.8	-0.044	762.41	-0.019
	511.31	-0.051	511.31	3	K	11	1410.7	-0.047	705.86	+0.0931
	639.37	+0.0498	639.37	4	Q	10	1168.6	-0.081	1168.6	
	738.44	+0.0144	738.44	5	V	9	1040.5	-0.037	1040.5	
	875.5	+0.0303	875.5	6	H	8	941.47	-0.015	941.47	
	972.55		972.55	7	P	7	804.41	+0.0137	804.41	
	1087.6	-0.056	1087.6	8	D	6	707.36	-0.013	707.36	
-0.415	594.82	-0.119	1188.6	9	T	5	592.33	-0.059	592.33	
	1245.6	-0.072	1245.6	10	G	4	491.28	+0.0045	491.28	
	1358.7	-0.121	1358.7	11	I	3	434.26		434.26	
	1445.8	-0.003	1445.8	12	S	2	321.18	+0.1483	321.18	
	1532.8	-0.003	1532.8	13	S	1	234.14		234.14	
				14	K	0	147.11		147.11	

H2BK108 $_{pr}$LLLPGELAK$_{glu}$HAVSEGTK$_{pr}$

Charge: 3 m/z: 995.0488

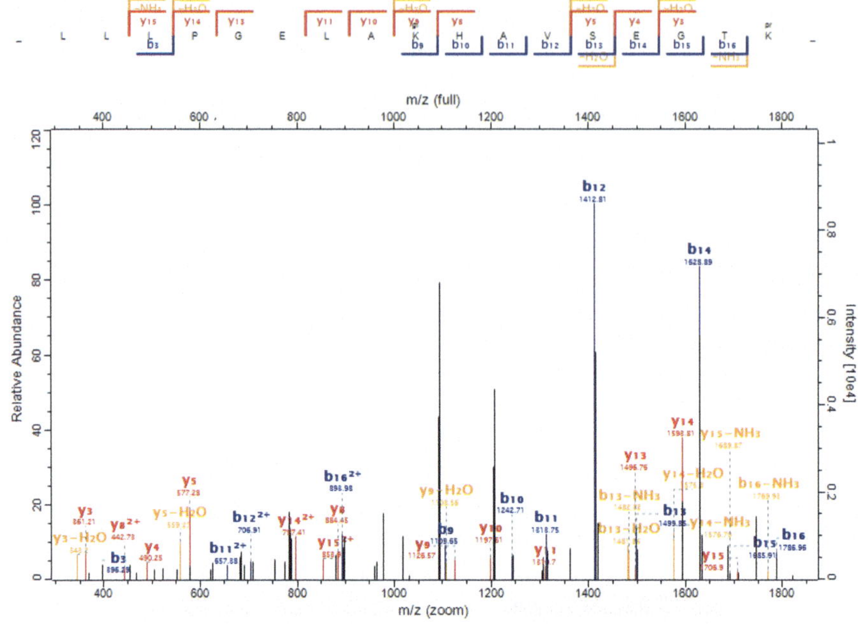

b²⁺ ion		b ion		seq			y ion		y²⁺ ion	
Δ dalton	mass	Δ dalton	mass				mass	Δ dalton	mass	Δ dalton
	170.12		170.12	1	L	16				
	283.2		283.2	2	L	15	1820		1820	
	396.29	+0.0788	396.29	3	L	14	1706.9	+0.0677	853.95	+0.2208
	493.34		493.34	4	P	13	1593.8	-0.013	797.41	+0.2126
	550.36		550.36	5	G	12	1496.8	+0.0155	1496.8	
	679.4		679.4	6	E	11	1439.7		1439.7	
	792.49		792.49	7	L	10	1310.7	+0.0549	1310.7	
	863.52		863.52	8	A	9	1197.6	-0.001	1197.6	
	1105.7	+0.1139	1105.7	9	K	8	1126.6	+0.0074	1126.6	
	1242.7	+0.0527	1242.7	10	H	7	884.45	+0.0589	442.73	-0.304
+0.3533	657.38	-0.002	1313.7	11	A	6	747.39		747.39	
+0.2358	706.91	+0.0293	1412.8	12	V	5	676.35		676.35	
	1499.8	+0.0436	1499.8	13	S	4	577.28	+0.0228	577.28	
	1628.9	+0.0137	1628.9	14	E	3	490.25	+0.1057	490.25	
	1685.9	+0.0349	1685.9	15	G	2	361.21	+0.0136	361.21	
+0.1697	893.98	+0.1115	1787	16	T	1	304.19		304.19	
				17	K	0	203.14		203.14	

H2BK116 $_{pr}$HAVSEGTK$_{glu}$AVTK$_{pr}$

Charge: 2 m/z: 727.3803

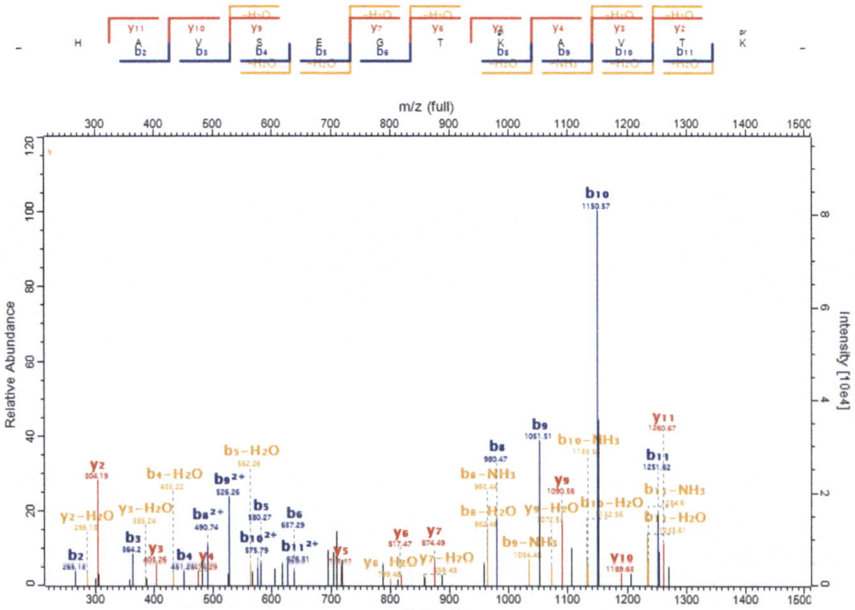

b²⁺ ion		b ion					y ion	
Δ dalton	mass	Δ dalton	mass		seq		mass	Δ dalton
	194.092		194.092	1	H	11		
	265.13	+0.13031	265.13	2	A	10	1260.67	-0.0622
	364.198	+0.05936	364.198	3	V	9	1189.63	-0.0389
	451.23	+0.11294	451.23	4	S	8	1090.56	-0.0076
	580.273	-0.0191	580.273	5	E	7	1003.53	
	637.294	+0.06469	637.294	6	G	6	874.488	+0.07136
	738.342		738.342	7	T	5	817.467	+0.17413
-0.0011	490.738	-0.0348	980.468	8	K	4	716.419	-0.0003
+0.05936	526.256	+0.04641	1051.51	9	A	3	474.292	+0.02937
+0.29866	575.791	-0.007	1150.57	10	V	2	403.255	+0.12804
+0.17393	626.314	-0.0812	1251.62	11	T	1	304.187	+0.11012
				12	K	0	203.139	

H2BK120 $_{pr}$AVTK$_{glu}$YTSAK$_{pr}$

Charge: 2 m/z: 597.8163

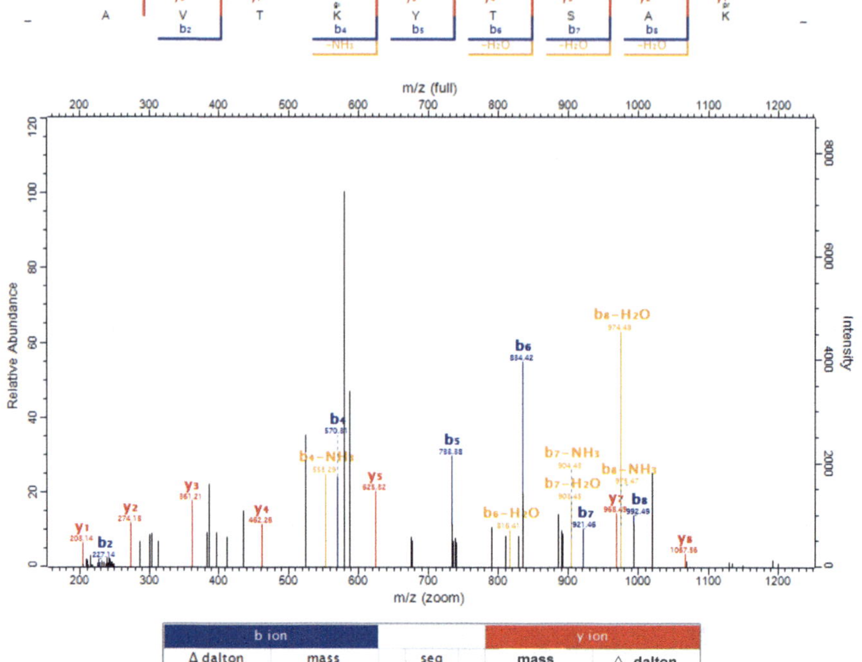

b ion					y ion	
Δ dalton	mass		seq		mass	△ dalton
	128.070605	1	A	8		
+0.080433	227.1390189	2	V	7	1067.561918	-0.0229772
	328.1866974	3	T	6	968.4935037	+0.0675925
-0.0079956	570.3133545	4	K	5	867.4458252	
-0.020848	733.376683	5	Y	4	625.3191681	-0.1355744
+0.0509193	834.4243615	6	T	3	462.2558396	-0.0146897
-0.0767512	921.4563899	7	S	2	361.2081611	+0.0917656
+0.0357932	992.4935037	8	A	1	274.1761327	+0.2338099
		9	K	0	203.1390189	+0.0354043

H2AK95 $_{pr}$NDEELNK$_{glu}$LLGR

Charge: 2 m/z: 735.8754

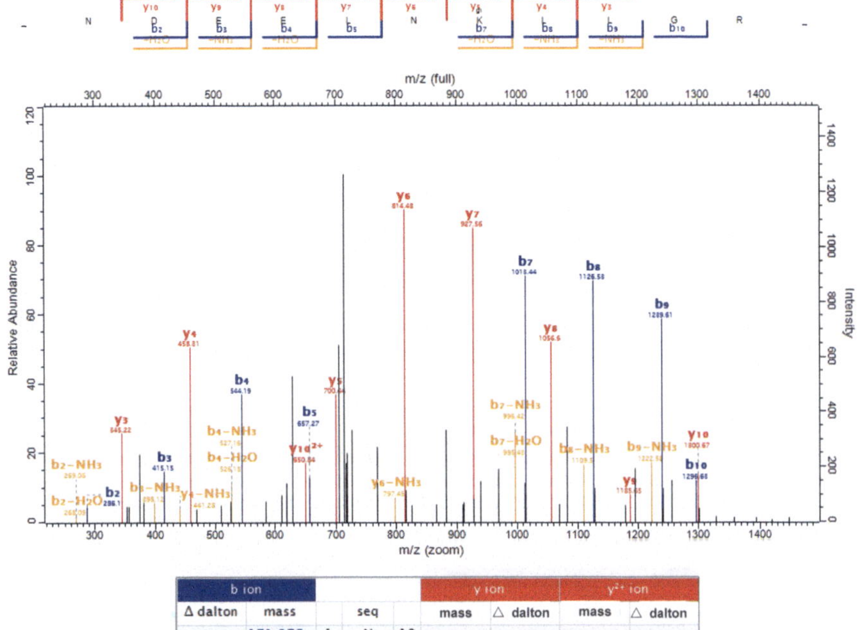

b ion					y ion		y^{2+} ion	
Δ dalton	mass		seq		mass	△ dalton	mass	△ dalton
	171.076	1	N	10				
+0.16232	286.103	2	D	9	1300.67	+0.04907	650.841	+0.01028
+0.04393	415.146	3	E	8	1185.65	+0.11666	1185.65	
-0.0336	544.189	4	E	7	1056.6	+0.10212	1056.6	
+0.03696	657.273	5	L	6	927.562	+0.09778	927.562	
	771.316	6	N	5	814.478	+0.08809	814.478	
+0.06372	1013.44	7	K	4	700.435	+0.04246	700.435	
+0.06175	1126.53	8	L	3	458.309	+0.04644	458.309	
+0.00174	1239.61	9	L	2	345.224	+0.10206	345.224	
+0.09294	1296.63	10	G	1	232.14		232.14	
		11	R	0	175.119		175.119	

H2AK99 LLGK$_{glu}$VTIAQGGVLPNIQAVLLPK

Charge: 3 m/z: 819.4999

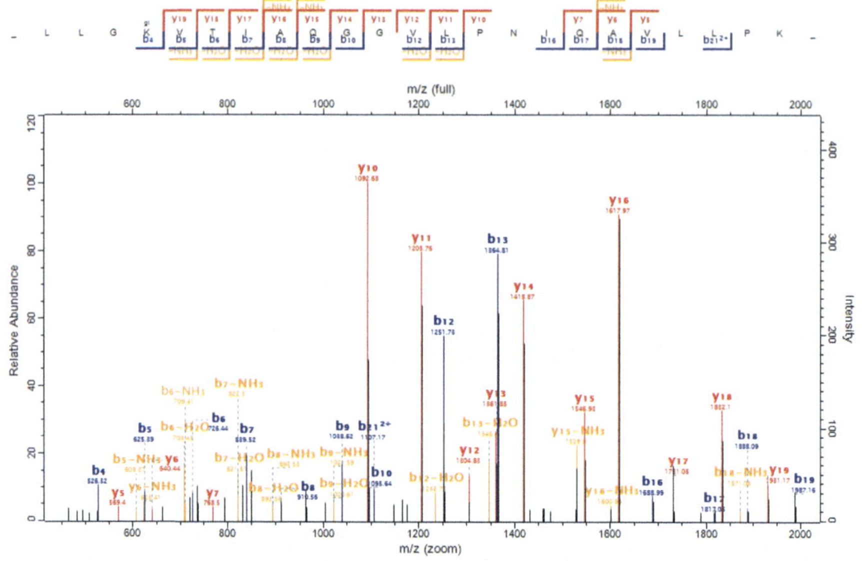

b²⁺ ion		b ion					y ion	
Δ dalton	mass	Δ dalton	mass		seq		mass	Δ dalton
	114.0913		114.0913	1	L	22		
	227.1754		227.1754	2	L	21	2343.401	
	284.1969		284.1969	3	G	20	2230.317	
	526.3235	-0.05033	526.3235	4	K	19	2173.295	
	625.3919	+0.059538	625.3919	5	V	18	1931.169	-0.00011
	726.4396	+0.003742	726.4396	6	T	17	1832.1	+0.002754
	839.5237	+0.112915	839.5237	7	I	16	1731.053	+0.070452
	910.5608	+0.102413	910.5608	8	A	15	1617.969	-0.06582
	1038.619	+0.135022	1038.619	9	Q	14	1546.932	-0.126
	1095.641	+0.061922	1095.641	10	G	13	1418.873	-0.06351
	1152.662		1152.662	11	G	12	1361.851	-0.11151
	1251.731	-0.05066	1251.731	12	V	11	1304.83	-0.01766
	1364.815	-0.07674	1364.815	13	L	10	1205.762	-0.04263
	1461.868		1461.868	14	P	9	1092.678	-0.0124
	1575.91		1575.91	15	N	8	995.6248	
	1688.995	+0.01218	1688.995	16	I	7	881.5819	
	1817.053	-0.01136	1817.053	17	Q	6	768.4978	+0.161134
	1888.09	-0.03334	1888.09	18	A	5	640.4392	+0.059678
	1987.159	-0.01264	1987.159	19	V	4	569.4021	+0.084462
	2100.243		2100.243	20	L	3	470.3337	
+0.170625	1107.167		2213.327	21	L	2	357.2496	
	2310.38		2310.38	22	P	1	244.1656	
				23	K	0	147.1128	

H2AK118 $_{pr}$VTIAQGGVLPNIQAVLLPK$_{glu}$K$_{pr}$

Charge: 3 m/z: 819.4999

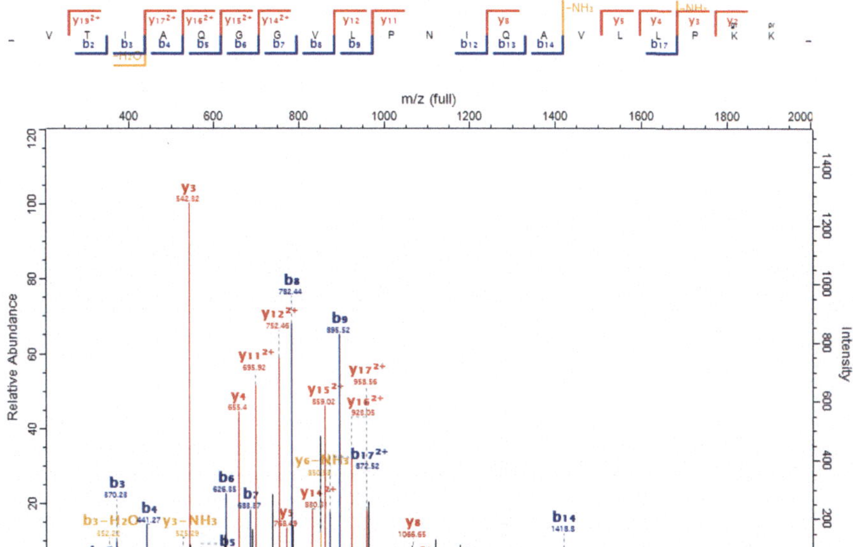

b²⁺ ion		b ion					y ion		y²⁺ ion	
Δ dalton	mass	Δ dalton	mass		seq		mass	△ dalton	mass	△ dalton
	156.1		156.1	1	V	19				
	257.15	+0.2029	257.15	2	T	18	2130.3		1065.6	+0.1416
	370.23	+0.1252	370.23	3	I	17	2029.2		2029.2	
	441.27	+0.0312	441.27	4	A	16	1916.1		958.56	+0.1323
	569.33	+0.0393	569.33	5	Q	15	1845.1		923.05	+0.2964
	626.35	+0.0525	626.35	6	G	14	1717		859.02	+0.214
+0.1789	342.19	+0.0635	683.37	7	G	13	1660		830.51	+0.2595
	782.44	+0.0624	782.44	8	V	12	1603		1603	
	895.52	+0.1075	895.52	9	L	11	1503.9	+0.1848	752.46	+0.1408
	992.58		992.58	10	P	10	1390.8	+0.0568	695.92	+0.1516
	1106.6		1106.6	11	N	9	1293.8		1293.8	
	1219.7	+0.0404	1219.7	12	I	8	1179.7		1179.7	
	1347.8	-0.014	1347.8	13	Q	7	1066.7	+0.0609	1066.7	
	1418.8	+0.0842	1418.8	14	A	6	938.59		938.59	
	1517.9		1517.9	15	V	5	867.55		867.55	
	1631		1631	16	L	4	768.49	+0.1382	768.49	
+0.1343	872.52	+0.2682	1744	17	L	3	655.4	+0.0965	655.4	
	1841.1		1841.1	18	P	2	542.32	-0.018	542.32	
	2083.2		2083.2	19	K	1	445.27	+0.1379	445.27	
				20	K	0	203.14		203.14	

H2AK119 $_{pr}K_{glu}$TESHHK$_{pr}$

Charge: 2 m/z: 546.7697

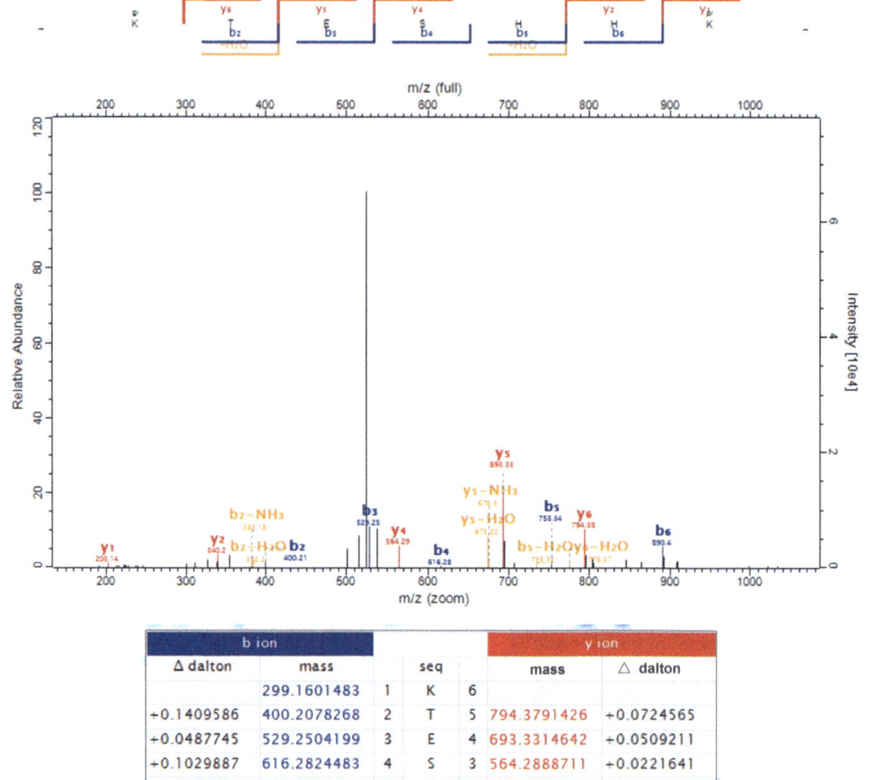

b ion					y ion	
Δ dalton	mass		seq		mass	△ dalton
	299.1601483	1	K	6		
+0.1409586	400.2078268	2	T	5	794.3791426	+0.0724565
+0.0487745	529.2504199	3	E	4	693.3314642	+0.0509211
+0.1029887	616.2824483	4	S	3	564.2888711	+0.0221641
+0.0474338	753.3413601	5	H	2	477.2568426	
+0.0765346	890.400272	6	H	1	340.1979308	+0.1619325
		7	K	0	203.1390189	-0.0168113

H2AK125 $_{pr}K_{pr}TESHHK_{glu}AK_{pr}$
Charge: 2 m/z: 674.3488

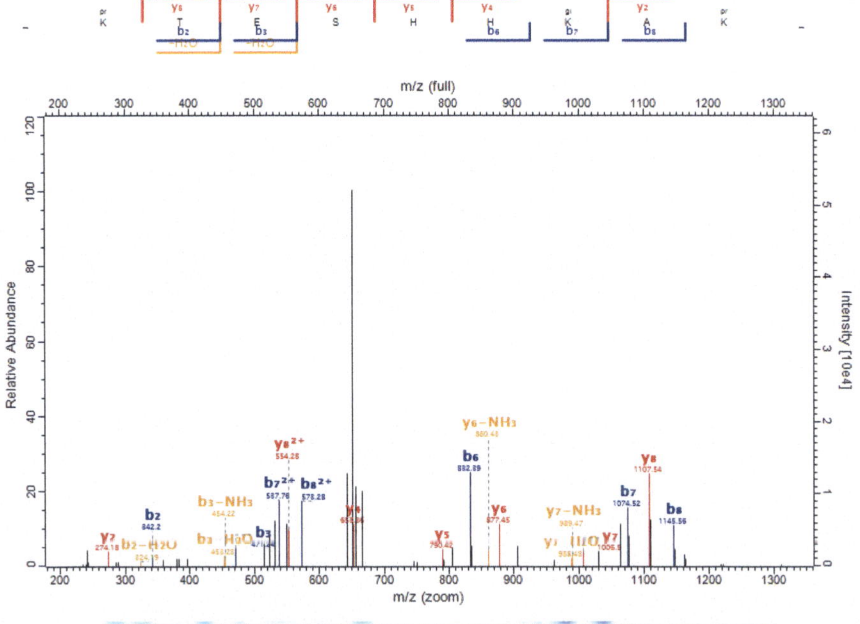

b²⁺ ion		b ion					y ion		y²⁺ ion	
Δ dalton	mass	Δ dalton	mass		seq		mass	△ dalton	mass	△ dalton
	241.15		241.15	1	K	8				
	342.2	+0.162	342.2	2	T	7	1107.5	+0.0442	554.28	+0.0887
	471.24	+0.1002	471.24	3	E	6	1006.5	+0.043	1006.5	
	558.28		558.28	4	S	5	877.45	+0.1087	877.45	
	695.34		695.34	5	H	4	790.42	+0.0959	790.42	
	832.39	+0.0749	832.39	6	H	3	653.36	-0.163	653.36	
+0.1723	537.76	+0.022	1074.5	7	K	2	516.3		516.3	
+0.1469	573.28	-0.043	1145.6	8	A	1	274.18	+0.2186	274.18	
				9	K	0	203.14		203.14	

The manufacturer's authorised representative in the EU is Springer Nature Customer Service Centre GmbH, Europaplatz 3, 69115 Heidelberg, Germany. If you have any concerns regarding our products, please contact ProductSafety@springernature.com

Printed and bound by CPI Group (UK) Ltd, Croydon, CR0 4YY
15/07/2026
02167645-0013